Simulation of ecologic

Simulation Monographs

Simulation monographs is a series on computer simulation in agriculture and its supporting sciences

Simulation of ecological processes

C.T. de Wit and J. Goudriaan

Wageningen
Centre for Agricultural Publishing and Documentation
1974

Books already published in this series:

C. T. de Wit and H. van Keulen
Simulation of transport processes in soils
1972, 108 pages, ISBN 90 220 0417 1, price Dfl. 15.00

J. Beek and M. J. Frissel
Simulation of nitrogen behaviour in soils
1973, 76 pages, ISBN 90 220 0440 6, price Dfl. 12.50

ISBN 90 220 0496 1

Cover design: Pudoc, Wageningen

Printed in Belgium

Contents

This paper is the result of work of students and research workers, interested in the simulation of ecosystems.

These are:

Agricultural University of Wageningen:
Ms. J. G. Blijenburg, J. Goudriaan, G. J. van Hoof, C. de Jonge, C. T. de Wit

University of Nijmegen:
J. G. M. Janssen (Chapter 6)

Hebrew University, Jerusalem (Israel):
Graduate students and research workers of the workshop 'Simulation in Ecology', organized by the late N. H. Tadmor (April 1973)

Connecticut Agricultural Experiment Station, New Haven (USA):
P. E. Waggoner (Chapter 7)

1 Introduction

1.1 Purpose

'In recent years there has been extensive study of the behaviour of complex interacting systems in such fields as engineering, physiology and economics. Drawing on and building upon this diverse body of experience, progress has been made over the past ten years in the development of methods for understanding the dynamics of ecosystems and the impact of stresses upon them, including stresses generated by man. These methods are subsumed under the heading of systems ecology. Systems ecology is based on the assumption that the state of an ecosystem at any particular time can be expressed quantitatively and that changes in the system can be described by mathematical terms,' to quote the expert panel on the role of system analyses of the Man and Biosphere Program (MAB, 1972).

Whether this basic assumption of systems ecology can be made operative or not, the approach raises considerable interest among natural scientists. However, it appears that many outsiders venturing in systems ecology are confused because they are exposed to mathematical and computer techniques, and to the treatment of complex systems at a too rapid rate. This is a pity because the systems approach has its merits. However, these can only come to the fore if a dialogue can be maintained between system ecologists and their more experimentally inclined colleagues.

The confusion may be reduced and the necessary dialogue stimulated by introducing the motivated ecologist stepwise to one of the main aspects of system ecology: the analyses and simulation of state determined systems. This is done in this book by treating in detail various ecological systems, ranging from simple exponential growth to a plant epidemic of considerable complexity. Ecological, mathematical and programming aspects are interwoven during the treatment, and exercises have been set that are an integral part of the text on a second reading. In this way, it appears that the only thing that may be new in simulation is the emphasis that is placed on the quantifying of the underlying processes, the iterative use of information and the use of suitable 'simulation languages' as a means of communication, not only between man and machine, but also between man and man.

1.2 Some terminology

System ecologists use and misuse many words and terms and in order not to add to the confusion of the reader it is necessary to define the common concepts: model, system and simulation.

There are many models. A simple mathematical model is the age-old relation of velocity (v) and the distance (s) covered by a falling apple, depending on the gravitational acceleration (g) and the time from the moment of release (t): $v=gt$ and $s=0.5\,vt^2$. An example of a non-mathematical model is a map. It is a simplification of the original that contains relevant information and allows measurements. Dependent upon the purpose of the map railways, lines of equal rainfall or soil types are presented.

A scale model of a ship enables measurement of the resistance of the original in the water. To maintain the original relations between viscosity, density, velocity, length etc., the laws of scaling must be satisfied. Of course, the internal structure of the ship is not modelled.

A system may be defined as a limited part of reality with related elements. The totality of relations within a system is called the structure of a system: both models and systems have a structure and it follows from the definitions that a model is system. The reverse does not seem true. However, it may be argued that a piece of art is a model of a conception in the artist's mind or that an engine is a model of the conception of its creator. A system M is a model of system O, provided that the structure is partly overlapping or isomorphic. Which parts of O are presented in M is determined by the requirements of relevance imposed on the model. Which part should not be considered follows also from the requirement that a model must remain easy to handle and lucid.

Examples of a system are a cell, a plant, an animal, a field with a crop, a forest and a farm. It is better to choose the boundary between system and environment such that the system is isolated. This is often not possible and then the boundary should be chosen so that the environment influences the system, but the system itself does not affect the environment. To achieve this goal it is often necessary to consider a larger system than seems necessary for the purpose. If for instance, the influence of temperature on the growth of plants is studied in a climate room, this climate room is part of the environment when its construction is so good that temperature, moisture content and light intensity do not depend on the size of the plants. In most climate rooms this requirement is not met so that it may be wise to treat the room itself as part of the system. If this leads to unwieldy models, it may be necessary to characterize the interaction between plant system and environment by continuous measurements at the interface, for example, of the light intensity at the leaf surfaces, and of the temperature and humidity between the plants. This approach erodes the generality of the model, but may enable better evaluation.

A system has a pattern of behaviour which implies that the system changes with time, that it is dynamic. A simplified representation of a dynamic system is a dynamic model. An operational definition of simulation is the building of a dynamic model and the study of its behaviour. Simulation is useful if it increases one's insight of reality by extrapolation and analogy, if it leads to the design of new experiments and if the model accounts for most relevant phenomena and contains no assumptions that are proven to be false. The latter requirement seems obvious, but is nevertheless formulated because such assumptions are often made to enable analytical solutions of mathematical models. With more recent simulation techniques this limitation can often be overcome, so that attention may be shifted from solution techniques to the study of behaviour of model and system.

There has been a tendency among statisticians to restrict the term simulation to the study and modelling of stochastic processes. However, this denies the use of the term in the field of the engineering sciences and what is even more unacceptable, it restricts too much the common usage of the word.

1.3 Electrical analogues

Many systems may be modelled by means of electrical analogues. For instance, a model of a falling apple might consist of an apparatus with two capacitors. The first one is charged with a current that is considered analogous with the gravitational acceleration and thus its potential is analogous with velocity. The second capacitor is charged with a current that is proportional to the potential of the first and thus its potential is analogous to the covered distance.

Exercise 1
Write down the differential equations that relate the rate of change of the velocity (v) of a falling apple to the gravitational acceleration (g) and of its distance (s) from the point of release to the velocity. Also write down the differential equation that relates the potential of a capacitor (c) to its capacitance and the charging current (i). Find the expressions for the charging currents i_1 and i_2 of two capacitors so that the potential of the first capacitor corresponds to v and of the second capacitor to s.

The integration in the capacitors takes place continuously and simultaneously, as do velocity and distance in reality. At any moment the condition of the system is fully determined by the potentials of the capacitors. The analogous computers that have been developed from this principle are very useful for simulating such continuous systems. There are, however, several problems with their use. The user should adapt the scale of variables to be modelled to the range of potential of the circuit elements and has to accept their inaccuracies. The resulting difficulties rapidly increase with increasing size and complexity of the models and with preciser requirements of accuracy.
These problems do not show up in simulation with digital computers and it is very illustrative that the first 'languages' to simulate continuous systems on digital computers were developed to control the result of analogous machines and to facilitate the assessment of the scaling factors.
Digital machines with proper simulation languages are also preferred to analogous machines when there are discontinuous elements, and when many empirical relationships are used. A large drawback, however, is that digital machines operate sequentially and discretely, whereas in many systems, continuous processes operate in parallel.
It is likely that in due course the disadvantages of both machines will be eliminated and the advantages combined by hybrid computers in which digital and analogous computers are amalgamated.

1.4 State determined systems

As has been said, systems ecology is based on the assumption that the state of an ecosystem at any particular time can be expressed quantitatively and that changes in the system can be described by mathematical terms. This assumption leads to the formulation of state determined models in which state variables, driving or forcing variables, rate variables, auxiliary variables and output variables can be distinguished.

State variables characterize and quantify all conserved properties of the system, such as amount of biomass, number of animals, content of mineral elements in various parts of the system, amount of food, amount of poison, number of niches, water content, temperature of the soil and so on. The values of all state variables have to be known at the onset of simulation. In mathematical terms they are quantified by the contents of integrals.

Driving or forcing variables are those that are not affected by processes within the system but characterize the influence from outside. These may be macrometeorological variables, the amount of food added in course of time and so on. It should be realized that depending on the boundaries of the system to be simulated, the same variables may be classified either as state or driving variables.

Rate variables quantify the rate of change of the state variables. Their values are determined by the state variables and the driving variables according to rules formulated from the knowledge of the underlying ecological, physiological and physical processes. These processes may be so complicated that the calculation process becomes much more lucid when use is made of proper chosen intermediate or auxiliary variables.

Output variables are the quantities which the model produces for the user. Sometimes they are state variables, sometimes rates and sometimes auxiliary variables that may be calculated especially for the purpose.

In such state determined models rates are not mutually dependent: each rate of change depends at any particular time on the values of the state and driving variables and can therefore be calculated independently of all other rates. Thus structural equations, that means n equations with n unknown rates, do not occur.

This is in accordance with experience. In a mixture of yeasts, the rates of growth do not depend on each other, but each one separately upon the state of the system, characterized by its own amount, the concentration of food and waste products and forcing variables such as temperature. The interaction between the growth of the yeasts in the mixture evolves in time because of the consumption of the same food source or production of the same waste products. Another example. In a chemical reaction where compound C is formed from compounds A and B, the rate of formation of C does not depend on the rate of formation of A and B but only on the amounts or concentrations of compound A, B and C that are present and the reaction constants. If, however, the rate of formation of A is slow, this compound is depleted in due course to such a level that the rate of formation of C adjusts to that of A. In other words, the observation that the rate of formation of A and C is equal is a consequence of the operation of the system and does not reflect a direct relation between both rates, that is to be modelled. Another example is presented in the form of an exercise.

Exercise 2

Ask two children, who are not allowed to communicate with each other, to stand on one side of a room and tell them to walk to the other side of the room at the same speed, carrying out independently of each other the following instructions on command:
1: close eyes, 2: decide on step size, 3: take a step, 4: open eyes, 5: compare positions, 1: close eyes, 2: decide on step size, and so on. Mark the position of each child on the floor after each cycle of instructions. Do the children stay practically side by side? What are the variables of state? What are their 'rates' of change? In how many independent decision processes are these decided upon? It may be observed that the subsequent step sizes of each child are more variable in the beginning than at the end. Why is this so?

1.5 Simulation languages

Although digital computers may memorize data easily, these sequential and digital instruments seem most unsuitable for the simulation of continuous and parallel dynamic systems. The main feature of simulation languages, or simulation programming systems, is to

overcome these disadvantages.
The principle that rates of change are not mutually dependent but depend independently on variables of state and driving variables, allows all rates of change to be calculated in any order at any instant of time. After calculating all rates at one instant, the changes are done by integrating the state variables over a small time-interval. In this way, the model operates in semi-parallel fashion. The simplest way of integration is by the Eulerian or rectilinear method in which the new value of a state variable at time $t+\Delta t$ equals the old value at time t plus the calculated rate of change at time t times the constant time-interval Δt.
Another feature of simulation programming systems is that all processes and processing details may be presented in conceptional rather than computational order. The programming system itself contains a sorting routine which orders all calculations and integrations in a proper algorithm. The advantages of this procedure are that the simulation program may be presented much more clearly, that a considerable variety of programming and conceptional errors may be detected by the system and that sub-models are easily assembled in a larger model. Apart from this, all simulation languages contain subroutines that execute operations that are often dealt with in modelling and facilitate the organization of data input and output.
In this way programming systems have been developed that enable sophisticated use of computers by research workers without much training in advanced programming techniques and with a minimum of formal mathematical knowledge. As has been said, these programming systems are not only intended to improve the communication between man and computer but also between research workers themselves. Since many simulation programs that appear at present can be used only by the designer himself or by very motivated programmers, this latter aspect of simulation languages is very important and needs to be developed still more.
Unfortunately, simulation programming systems have proliferated to such an extent during the last 12 years that this latter purpose is continuously threatened. However, recently all languages seem to be gravitating towards one concept: the Continuous System Simulation Language (CSSL), originally defined by a working party of the Simulation Council (Brennan and Silberberg, 1968). The most widely

used CSSL version is at present the Continuous System Modeling Program (IBM). This language as defined in IBM Users Manual SH20–0367–4 is used in the subsequent text, together with a preprocessor that facilitates the simulation of systems that vary in time and space and the handling of historical information. This preprocessor is available on request from the computer centre of the Agricultural University in Wageningen. This centre also provides DSLH (Digitale Simulatie Landbouw Hogeschool), a continuous simulation language developed from DSL for use on CDC computers with a capacity of 32 K upwards. The present text is self-explanatory and written to be read without the use of a manual, but for actual programming it is advisable to have a manual at hand.

2 Exponential growth

2.1 Analytical and numerical integration

The growth rate of many populations may be proportional to the size of the population, either expressed in number of individuals or total biomass. With simple organisms such as bacteria, growth is often also continuous. The growth rate at any moment can then be expressed by the equation:

$$GR = RGR \times A \qquad (2.1)$$

in which RGR is the relative growth rate and A the amount of organisms.

Exercise 3
If weight is expressed in grams and time in hours, what are then the dimensions of A, RGR and GR? Give at least three environmental conditions that must always be satisfied to achieve a situation where the relative growth rate is independent of the amount of organisms and time.

In differential notation, Eqn (2.1) is written as

$$dA/dT = RGR \times A \qquad (2.2)$$

The integrated form of this equation or the analytical solution when RGR is constant, is

$$A = IA \times e^{RGR \times T} \qquad (2.3)$$

in which IA is the amount of organisms that appear to be present at time zero and *e* the base of the natural logarithm. Under these circumstances the amount of organisms increases exponentially with time.

Exercise 4

Calculate with a slide rule, with tabulated values of the function e^x or with a table of logarithms, the value of A after 0, 2, 4 up to 10 hours for RGR equal to 0.1 hour^{-1} and IA equal to 1 gram. Represent the results by a graph with time along the horizontal axis and the amount A along the vertical axis and connect the points with a smooth line. Plot the results also on a graph with time along the horizontal axis and the logarithm of the amount A along the vertical axis and connect the points also by a line.

What do you observe about the straightness of the second line? Show that this observation is mathematically correct.

The amount of organisms as a function of time may be found also by a recursive process. If, at a certain time T the amount of organisms equals A, the rate of growth at that moment equals RGR × A. During a short time-interval delta time (DELT), this rate of growth hardly changes, so that at time T+DELT the amount of organisms equals approximately A+RGR × A × DELT. With this new value, the rate of growth at time T+DELT can be calculated and so the amount of organisms at time T+2 × DELT, and so on.

Exercise 5

Calculate the values of A after 0, 2, 4 up to 10 hours for RGR = 0.1 hour^{-1} and H = 1 gram at time zero. Use time intervals of 2 hours and apply the following scheme:

TIME	A	RGR × A	RGR × A × DELT
0	1	0.1	0.2
2	1.2	etc.	

Plot the results on the graphs of Exercise 4 and connect the points by straight line segments.

A comparison of this step-wise solution and the analytical solution shows that the size of the population is underestimated by the use of the recursive solution. This is caused by the wrong assumption that the growth rate remains the same during each time-interval DELT, even though for continuous growth, the amount of organisms increases. It is to be expected that the discrepancy between the recursive and analytical solution decreases with decreasing time-intervals.

Exercise 6

Plot the results of a recursive calculation for time steps of 1 and 0.5 hours on the graphs of Exercise 4. Derive a formula that gives the value of H directly after n time-intervals of size DELT and convert this function of n to a function of time.

2.2 Simulation

Calculations with even shorter time-intervals are very tedious and are better done by formulating the problem in CSMP and using a computer. In its elementary form, this simulation is the same as numerical integration of a set of differential equations.
The problem in CSMP reads as follows on punched cards:

```
TITLE EXPONENTIAL GROWTH
A = INTGRL (IA,GR)
GR = RGR * A
INCON IA = 1.
PARAMETER RGR = 0.1
TIMER FINTIM = 10., OUTDEL=0.5, DELT=0.1
PRTPLT A
METHOD RECT
END
STOP
```

The first card mentions the TITLE, which is repeated on every page of output. The card with the INTGRL function states that A equals IA at time zero and that its current value at any time is found by integrating GR. The fourth and fifth card give the value of the only INitial CONstant (IA) and of the only PARAMETER (RGR). The TIMER card ensures that the simulation is finished after 10 time-units (FINTIM), that output is given at every 0.5 time unit (OUTDEL) and that intervals of 0.1 unit (DELT) are used for the numerical integration.

Exercise 7
Which variable governs the unit of time?

It is stated on the PRinTPLoT card that the value of A has to be plotted against time and that its numerical value has to be given also in a table. The METHOD card indicates that the integration must be done according to the RECTilinear method of Euler. This is the method that was used in the previous exercises. The END card defines the end of the simulation model and the STOP card the end of the simulation program. If the computation is to be repeated with a relative growth rate of 0.2, it suffices to insert the cards PARAMETER RGR = 0.2 and END between the END and STOP cards in the above program.

Exercise 8
Punch the lines of the program on cards and urge your computer centre to install one of the CSSL-type languages and to inform you on the deviation in notation between this language and the CSMP version used in this example. Then carry out the program.

Some readers may not have access to a suitable computer centre so that a facsimile of program and results are given in Fig. 1. The first

Page 1

```
                ****CONTINUOUS SYSTEM MODELING PROGRAM****
          ***PROBLEM INPUT STATEMENTS***
      TITLE EXPONENTIAL GROWTH
            A=INTGRL(IA,GR)
            GR=RGR*A
      INCON IA=1.
      PARAMETER RGR=0.1
      TIMER FINTIM=10.,OUTDEL=0.5,DELT=0.1
      PRTPLT A
      METHOD RECT
      END
      STOP
OUTPUT VARIABLE SEQUENCE
GR     A
 OUPUTS   INPUTS    PARAMS  INTEGS + MEM BLKS  FORTRAN  DATA CDS
  6(500)  24(1400)  5(400)     1+   0=  1(300)    3(600)      7
```

Page 2

```
                ***CSMP/360 SIMULATION DATA***
TITLE EXPONENTIAL GROWTH
INCON IA=1.
PARAMETER RGR=0.1
TIMER FINTIM=10.,OUTDEL=0.5,DELT=0.1
PRTPLT A
METHOD RECT
END
```

Page 3

```
FORTRAN IV G LEVEL  21                    UPDATE                    DATE = 7319C
0001              SUBROUTINE UPDATE
0002              COMMON  ZZ9901(5),IZ9901,ZZ9902,IZ9902,ZZ9903,IZ9903,ZZ9991(54)
0003              COMMON TIME
                 1,DELT   ,DELMIN,FINTIM,PRDEL ,OUTDEL,A      ,GR     ,IA     ,RGR
0004              COMMON ZZ9992(7990),NALARM,IZ9993,ZZ9994(417),KEEP,ZZ9995(489)
                 f,IZ0000,ZZ9996(824),IZ9997,IZ9998,ZZ9999(  45)
0005              REAL        IA
0006              GO TO(39995,39996,39997,39998),IZ0000
            C    SYSTEM SEGMENT OF MODEL
0007        39995 CONTINUE
0008              IZ9993=  11
0009              IZ9997=   1
0010              IZ9998=  10
0011              READ(5,39990)(ZZ9999(IZ9999),IZ9999=1,   45)
0012        39990 FORMAT(18A4)
0013              IZ9901=    100010
0014              IZ9902=    100011
0015              IZ9903=        45
0016              GO TO 39999
            C    INITIAL SEGMENT OF MODEL
0017        39996 CONTINUE
0018              GO TO 39999
            C    DYNAMIC SEGMENT OF MODEL
0019        39997 CONTINUE
0020              GR=RGR*A
            C     A          =INTGRL     (IA        ,GR
0021              GO TO 39999
            C    TERMINAL SEGMENT OF MODEL
0022        39998 CONTINUE
0023        39999 CONTINUE
0024              RETURN
0025              END
```

Page 4

```
                          MINIMUM            A       VERSUS TIME                  MAXIMUM
                         1.0000E 00                                              2.7047E 00
 TIME              A             I                                                      I
0.0             1.0000E 00     +
5.0000E-01      1.0510E 00     -+
1.0000E 00      1.1046E 00     ---+
1.5000E 00      1.1610E 00     ----+
2.0000E 00      1.2202E 00     ------+
2.5000E 00      1.2824E 00     --------+
3.0000E 00      1.3478E 00     ----------+
3.5000E 00      1.4166E 00     ------------+
4.0000E 00      1.4888E 00     --------------+
4.5000E 00      1.5648E 00     ----------------+
5.0000E 00      1.6446E 00     ------------------+
5.5000E 00      1.7285E 00     ---------------------+
6.0000E 00      1.8167E 00     -----------------------+
6.5000E 00      1.9093E 00     --------------------------+
7.0000E 00      2.0067E 00     -----------------------------+
7.5000E 00      2.1091E 00     --------------------------------+
8.0000E 00      2.2167E 00     -----------------------------------+
8.5000E 00      2.3297E 00     ---------------------------------------+
9.0000E 00      2.4486E 00     ------------------------------------------+
9.5000E 00      2.5735E 00     ----------------------------------------------+
1.0000E 01      2.7047E 00     --------------------------------------------------+
```

Fig. 1 | A simulation program for exponential growth written in CSMP. Page 1 contains the punched program and the other pages are generated by the system. Page 2 contains parameters, constants and other numerical data. Page 3 the FORTRAN subroutine 'UPDATE' and Page 4 the answers.

page contains the program, and some additional information, the third page the FORTRAN subroutine 'UPDATE' created by CSMP and the fourth page the answers. The third page is mainly of interest for those readers that know some FORTRAN and wish to follow the way CSMP organizes the work. They may have also some use for several pages with information on the organization of the program that are not reproduced here.

Exercise 9

Plot the results also on the graph of Exercise 4 and compare the numerical results with those of the analytical solution. Explain why the simulated results are still underestimates.

For simple exponential growth where the relative growth rate is a constant, numerical integration or simulation is not necessary because the solution may be found analytically. However, the analytical approach is frustrated only by slight variations in the system.
For instance, the relative growth rate of a bacterial population may depend on temperature, so that in a series of experiments with a bacteria species the following observations of the relative growth rate could have been made:

TEMP (°C)	0	5	10	20	30	40	50
RGR (h^{-1})	0.0	0.04	0.07	0.17	0.19	0.26	0.25

Exercise 10

These are (faked) observational data. Represent the results on a graph and draw a smooth line through the data points.

In many situations, bacteria populations are not exposed to a constant temperature; the temperature varies more or less in daily cycles and the question may be posed what the growth rate is under such conditions. Obviously, the relative growth rate is then not a constant which may be defined on a parameter card but a variable which is some function of temperature. To simulate this situation the PARAMETER card which defines the relative growth rate is removed from the program of Fig. 1 and replaced by the following function statement:

```
RGR = AFGEN(RGRTB,TEMP)
```

This Arbitrary Function GENerator states that the value of RGR depends on the temperature (TEMP), according to a tabulated function with the table name RGRTB. This function is introduced into the simulation program in tabulated form on a FUNCTION card:

```
FUNCTION RGRTB=(0.,0),(10.,0.075),...
(20.,0.16),(30.,0.215),(40.,0.24),...
(50.,0.25)
```

The first number between each pair of brackets presents a value of the independent variable (TEMP) and the second one the corresponding value of the dependent variable (RGR); the three dots at the end of the first line indicate that the table is continued on the next line. The AFGEN function finds the value of the RGR at the current temperature by linear interpolation between the tabulated values: i.e. if TEMP equals 25°, then RGR equals $.16+(5/10)(.20-.16)=0.18$.

Exercise 11
Enter the tabulated data of FUNCTION RGRTB on the graph of Exercise 10 and join these data points by straight lines. This broken line now represents the relation between RGR and TEMP as introduced in the simulation model. Try to match your smoothed curve more satisfactorily by tabulating values for RGR at 2.5°C intervals. This does not seem worth the trouble here. Why not?

The next step is to define how the forcing variable TEMPerature varies with TIME. This may also be done with a function generator:

```
TEMP = AFGEN (TMPTB,TIME)
FUNCTION TMPTB = ..................
```

These tabulated functions of forcing variables tend to be very long because they have to cover the whole simulated time-span in sufficient detail. Often it suffices to present the experimental data by some mathematical function. For instance, if there is a daily temperature variation, a sinusoidal function may be used:

```
TEMP = AVTMP + AMPTMP*SIN(6.28*TIME/24.)
```

The function SIN calculates the sine value of the variable in the argument: 6.28 stands for $2\times\pi$, TIME is the simulated time in hours since the start of the simulation and 24 stands for the hours in a day. The average temperature and the amplitude of the temperature are given by

```
PARAMETER AVTMP = 20.,AMPTMP = 10.
```

Exercise 12
Prepare a graph that shows the course of temperature during 24 hours. At what time is the temperature at its maximum?

The variable TIME is always needed in dynamic models and the simulation language automatically keeps track of it.

Exercise 13
Reason that TIME could also be kept track of by the statement:

```
T = INTGRL(0.,1.)
```

A facsimile of the program and the output is given in Fig. 2. Note that the variables RGR and GR are entered on the PRTPLT card between brackets. This means that only the printed output of these variables is requested, not a graphical display. Note also that the 'UPDATE' contains all equations in computational order, whereas in the program itself they are presented in some conceptional order. It is obvious that the readability of such simple simulation programs does not depend very much on the sequence of the equations.

Fig. 2 | A simulation program for exponential growth with a temperature dependent relative growth rate.

```
                  ****CONTINUOUS SYSTEM MODELING PROGRAM****                Page 1
              ***PROBLEM INPUT STATEMENTS***
                 A=INTGRL(IA,GR)
                 GR=RGR*A
          INCON  IA=1.
                 RGR=AFGEN(RGRTB,TEMP)
          FUNCTION RGRTB=(0.,0.),(10.,0.075),(20.,0.16), ...
                        (30.,0.215),(40.,0.24),(50.,0.25)
          PARAMETER AVTMP=20.,AMPTMP=10.
                 TEMP=AVTMP+AMPTMP*SIN(6.28*TIME/24.)
          TIMER FINTIM=48.,OUTDEL=1.,DELT=0.5
          PRTPLT A(RGR,GR)
          METHOD RECT
          END
          STOP
OUTPUT VARIABLE SEQUENCE
TEMP    RGR     GR      A
 OUTPUTS     INPUTS      PARAMS     INTEGS + MEM BLKS     FORTRAN     DATA CDS
  8(500)    29(1400)     7(400)      1+   0=   1(300)      5(600)         8
```

Page 2

```
FORTRAN IV G LEVEL  21                    UPDATE        DATE = 73171         19/11/18
 0001             SUBROUTINE UPDATE
 0002             COMMON  ZZ9901(5),IZ9901,ZZ9902,IZ9902,ZZ9903,IZ9903,ZZ9991(54)
 0003             COMMON TIME
                 1,DELT   ,DELMIN,FINTIM,PRDEL ,OUTDEL,A      ,GR     ,IA     ,RGRTB
                 1,AVTMP  ,AMPTMP,RGR    ,TEMP
 0004             COMMON ZZ9992(7986),NALARM,IZ9993,ZZ9994(417),KEEP,ZZ9995(489)
                 f,IZ0000,ZZ9996(824),IZ9997,IZ9998,ZZ9999(  51)
 0005             REAL        IA
 0006             GO TO(39995,39996,39997,39998),IZ0000
             C    SYSTEM SEGMENT OF MODEL
 0007        39995 CONTINUE
 0008             IZ9993=  15
 0009             IZ9997=   1
 0010             IZ9998=  14
 0011             READ(5,39990)(ZZ9999(IZ9999),IZ9999=1,  51)
 0012        39990 FORMAT(18A4)
 0013             IZ9901=   120010
 0014             IZ9902=   140013
 0015             IZ9903=       51
 0016             GO TO 39999
             C   INITIAL SEGMENT OF MODEL
 0017        39996 CONTINUE
 0018             GO TO 39999
             C   DYNAMIC SEGMENT OF MODEL
 0019        39997 CONTINUE
 0020             TEMP=AVTMP+AMPTMP*SIN(6.28*TIME/24.)
 0021             RGR=AFGEN(RGRTB,TEMP)
 0022             GR=RGR*A
             C    A          =INTGRL      (IA           ,GR
 0023             GO TO 39999
             C  TERMINAL SEGMENT OF MODEL
 0024        39998 CONTINUE
 0025        39999 CONTINUE
 0026             RETURN
 0027             END
```

Page 3

```
                         MINIMUM             A      VERSUS TIME            MAXIMUM
                        1.0000E 00                                        1.0303E 03
 TIME              A            I                                              I            RGR           GR
0.0              1.0000E 00    +                                                        1.6000E-01    1.6000E-01
1.0000E 00       1.1703E 00    +                                                        1.7423E-01    2.0389E-01
2.0000E 00       1.3874E 00    +                                                        1.8749E-01    2.6012E-01
3.0000E 00       1.6642E 00    +                                                        1.9888E-01    3.3097E-01
4.0000E 00       2.0160E 00    +                                                        2.0762E-01    4.1855E-01
5.0000E 00       2.4598E 00    +                                                        2.1312E-01    5.2423E-01
6.0000E 00       3.0139E 00    +                                                        2.1500E-01    6.4798E-01
7.0000E 00       3.6959E 00    +                                                        2.1314E-01    7.8774E-01
8.0000E 00       4.5209E 00    +                                                        2.0766E-01    9.3882E-01
9.0000E 00       5.4985E 00    +                                                        1.9894E-01    1.0939E 00
1.0000E 01       6.6305E 00    +                                                        1.8756E-01    1.2436E 00
1.1000E 01       7.9090E 00    +                                                        1.7431E-01    1.3786E 00
1.2000E 01       9.3175E 00    +                                                        1.6009E-01    1.4916E 00
1.3000E 01       1.0813E 01    +                                                        1.3814E-01    1.4938E 00
1.4000E 01       1.2298E 01    +                                                        1.1764E-01    1.4467E 00
1.5000E 01       1.3727E 01    +                                                        1.0002E-01    1.3729E 00
1.6000E 01       1.5081E 01    +                                                        8.6478E-02    1.3042E 00
1.7000E 01       1.6374E 01    +                                                        7.7946E-02    1.2763E 00
1.8000E 01       1.7657E 01    +                                                        7.5000E-02    1.3243E 00
1.9000E 01       1.9012E 01    +                                                        7.7841E-02    1.4799E 00
2.0000E 01       2.0556E 01    +                                                        8.6275E-02    1.7735E 00
2.1000E 01       2.2434E 01    -+                                                       9.9728E-02    2.2373E 00
2.2000E 01       2.4825E 01    -+                                                       1.1729E-01    2.9116E 00
2.3000E 01       2.7953E 01    -+                                                       1.3775E-01    3.8505E 00
2.4000E 01       3.2099E 01    -+                                                       1.5973E-01    5.1271E 00
2.5000E 01       3.7556E 01    -+                                                       1.7406E-01    6.5370E 00
2.6000E 01       4.4517E 01    --+                                                      1.8734E-01    8.3396E 00
2.7000E 01       5.3393E 01    --+                                                      1.9875E-01    1.0612E 01
2.8000E 01       6.4672E 01    ---+                                                     2.0753E-01    1.3421E 01
2.9000E 01       7.8904E 01    ---+                                                     2.1307E-01    1.6812E 01
3.0000E 01       9.6674E 01    ----+                                                    2.1500E-01    2.0785E 01
3.1000E 01       1.1855E 02    -----+                                                   2.1318E-01    2.5274E 01
3.2000E 01       1.4502E 02    ------+                                                  2.0775E-01    3.0128E 01
3.3000E 01       1.7640E 02    --------+                                                1.9906E-01    3.5114E 01
3.4000E 01       2.1274E 02    ----------+                                              1.8771E-01    3.9934E 01
3.5000E 01       2.5380E 02    ------------+                                            1.7448E-01    4.4283E 01
3.6000E 01       2.9904E 02    --------------+                                          1.6026E-01    4.7925E 01
3.7000E 01       3.4712E 02    ----------------+                                        1.3840E-01    4.8042E 01
3.8000E 01       3.9486E 02    -------------------+                                     1.1787E-01    4.6543E 01
3.9000E 01       4.4084E 02    ---------------------+                                   1.0021E-01    4.4176E 01
4.0000E 01       4.8442E 02    -----------------------+                                 8.6614E-02    4.1957E 01
4.1000E 01       5.2603E 02    -------------------------+                               7.8017E-02    4.1039E 01
4.2000E 01       5.6726E 02    ---------------------------+                             7.5001E-02    4.2545E 01
4.3000E 01       6.1080E 02    -----------------------------+                           7.7772E-02    4.7503E 01
4.4000E 01       6.6034E 02    --------------------------------+                        8.6141E-02    5.6882E 01
4.5000E 01       7.2055E 02    ----------------------------------+                      9.9538E-02    7.1722E 01
4.6000E 01       7.9720E 02    --------------------------------------+                  1.1705E-01    9.3313E 01
4.7000E 01       8.9743E 02    -------------------------------------------+             1.3749E-01    1.2339E 02
4.8000E 01       1.0303E 03    -------------------------------------------------+       1.5946E-01    1.6429E 02
```

Exercise 14
Make reruns with a two times larger and a two times smaller value of DELT than used in the program. Is DELT = 0.5 hour a reasonable choice? Which values have to be entered on the parameter card to obtain results that are the same as those of the program in Fig. 2?

2.3 Time interval of integration

One of the main problems of such numerical integrations, whether done by hand or through the intermediate of a simulation language is the choice of the correct time-interval of integration.

Exercise 15
Construct a graph with the amount of organisms after 10 hours of growth against the magnitude of the time interval (DELT) ranging from 2 to 0.1 hours for RGR = 0.1 $hour^{-1}$. Enter the correct answer obtained by means of the analytical solution also in the same graph by a horizontal line.

One may argue that an acceptable time-interval gives an answer within a specified range from the answer obtained with the analytical solution. However true, this is not a very meaningful approach because it is unnecessary to find some, always approximate, answer by simulation when the correct analytical result is available.
Another approach is to assume that a correct time-interval is reached if halving its value does not change the relevant results of the simulation by more than a preset relative amount. A relative amount that must not be taken smaller than warranted by the accuracy of the parameters, initial constants and tabulated functions. Thus a reasonable compromise is obtained between the apparent accuracy of the answer and the computational effort which increases at least linearly with decreasing magnitude of the time interval. It must be realized, however, that the method is not foolproof, so that in other than trivial situations it is always necessary to evaluate the answers with common sense against a background of experience and experimental results.

Exercise 16
What is the largest acceptable time-interval in Exercise 15 when a relative apparent accuracy of 5 percent is acceptable? How much smaller or larger is this time-interval when RGR equals 0.2 instead of 0.1 $hour^{-1}$? Does the largest acceptable time-interval depend here on the length of the simulated period (FINTIM)? This is a difficult question!

The largest acceptable time-interval is practically the same throughout a simulation as long as the relative change of the state variables, in this case of the amount of organisms, remains the same. This situation exists in the first example, but in the second example the relative change of the amount of organisms is large during periods of high temperatures and small during periods of low temperatures. Hence it could be an advantage to adapt the size of the time interval during the simulation.
This is possible by using more sophisticated methods of integration. Several such methods are available in CSMP, but only that of Runge-Kutta/Simpson is a safe choice for simulating continuous ecosystems. This method is invoked by the card METHOD RKS. Then, instead of the rate at time T, a weighted rate at times T, T+DELT/2 and T+DELT is used in such a way that a correct integration is achieved within the time range, when the resulting integral function can be presented by a fourth-order polynomial. Whenever output of rates is requested, it is always the actual rate or slope of the integral at the printing time that is printed, not the average rate as used during this time interval. The size of a correct time-interval is obtained by comparing the result of this numerical integration with the result of a second order integration method of Simpson. If the error is too large, the time interval DELT is halved and if it is small enough, DELT is doubled for the next step. Hence the size of the time interval changes automatically to its largest acceptable value with varying relative rates of change. When the error criterion is not met by decreasing the time interval, the simulation is terminated. This often indicates a programming or conceptual error.
The procedure of calculating rates of change is therefore done several times before an integration is really executed. The advantages are a much larger time-interval of integration and a much larger accuracy,

although the latter is not always of great importance in ecological systems where negative feedback plays a role and the basic data are often not very accurate.
If the time interval does not change very much during simulation the method RKS uses a considerable fraction of the computer time trying again and again to increase the time interval of integration without success. Then it may be worthwhile to use METHOD RKSFX, a fourth order Runge-Kutta method with fixed time-interval. However, the time-interval has to be specified then by the user. An acceptable time-interval is found by executing the program once with the method RKS and counting the number of integrations.
Since

$$\mathrm{COUNT}_{\mathrm{T+DELT}} = \mathrm{COUNT} + (1/\mathrm{DELT}) \times \mathrm{DELT} = \mathrm{COUNT}_{\mathrm{T}} + 1$$

the statement

```
COUNT = INTGRL(0,1/DELT)
```

seems to count this number. However, this simple method cannot be applied when variable time-interval methods of integration are used because the rate of change (1/DELT) increases then in proportion with a decrease in DELT, so that shortening the time-interval does not decrease the relative error of this integration. It is therefore necessary to compute the number of integrations in such a way that this computation does not interfere with the integration process. This can be done by adding just above the END statement of the program a so called NOSORT section to execute some FORTRAN compilations. The statements in such a section are presented in computational order and bypassed by the sorting routine.
A suitable section would be:

```
NOSORT
IF(TIME.EQ.0.) COUNT=0.
COUNT=COUNT + KEEP
```

The first statement sets a counter at zero when TIME equals zero and the second statement adds the value of KEEP to COUNT every time the rate calculations are performed. However, the value KEEP is an internal CSMP variable which is set to 1 when the actual updating is done after executing all iterations demanded by the integration method, and set to zero otherwise.

Exercise 17
Add the NOSORT section to the program of Fig. 2 and do not forget to enter COUNT on the PRTPLT card. Carry out the program with METHOD RKS. It is then found that COUNT=FINTIM/OUTDEL. Explain why. What DELT should be specified when using the method RKSFX?
Execute the program also for ten times larger values of RGR and tabulate the average RGR and the number of counts, both between successive intervals of output. Introduce for this purpose the statement

```
RGRS = INTGRL(0.,RGR)
```

Estimate a maximum DELT for use with the method RKSFX.

One may ask why rectilinear integration is still used when other integration methods are so much satisfactory. However, sophisticated integration methods can be used in general only when all changes are continuous. In ecological models this is often not so; sudden changes in environmental conditions, sudden immigration and emigration and sudden death and other accidents may take place. It will be shown in Section 5.2 that such discontinuities can be programmed, but that then the most unsophisticated, rectilinear method of integration must be used.
Of course much more can be said about the use and misuse that can be made of numerical methods of integration, but this goes beyond the scope of this monograph and the more so because the technique is described in many good handbooks (e.g. Milne, 1960).

3 The growth of yeast

3.1 Description of the system

Growth is only exponential as long as the relative growth rate remains constant. This is usually so with yeast when it is grown under aerobic conditions with a sufficient supply of sugar and some other growth essentials. The sugar is then continuously consumed to provide the 'C skeletons' and the energy both for the growth of new yeast cells and for maintenance of the yeast. The end-products, CO_2 and H_2O, of the sugar broken down in the respiratory process do not pollute the environment of the yeast.
However, if yeast grows under anaerobic conditions, one end-product of the respiratory processes is alcohol which may accumulate in the environment. This slows down and ultimately stops the development of yeast buds even when there is still enough sugar available for growth.
Growth curves for yeast that result under such conditions are given in Fig. 3. It should be noted that yeast once formed remains because only the bud formation is affected by the alcohol; the yeast itself is not killed. Two of the four growth curves are from an experiment of Gause (1934) with monocultures of the yeast species *Saccharomyces cerevisiae* and *Schizosaccharomyces* '*Kephir*'. It is obvious that the initial relative growth rate and the maximum volume of yeast that is ultimately formed is highest for the first species.
Gause cultivated both yeast species not only in monoculture, but also in mixture. The results of this experiment are also presented in Fig. 3 by the other two curves. A comparison of the growth of both species in mixture with their growth in monoculture shows that both affected each other in the first situation. It was proposed by Gause that this was due to the formation of the same waste product, alcohol, that affected the bud formation of both species. In this chapter we shall analyse whether this explanation is acceptable by constructing a model that simulates the growth of two species independently and in mixture

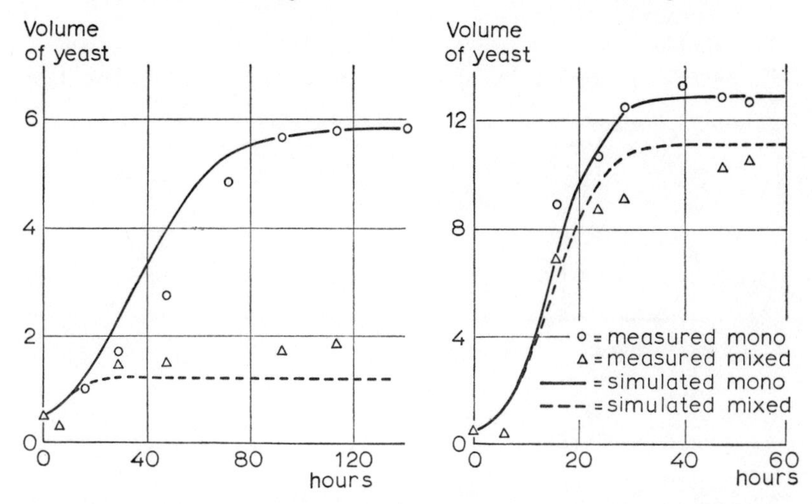

Fig. 3 | The growth of *Saccharomyces cerevisiae* and *Schizosaccharomyces 'Kephir'* in monoculture and in mixture. The observational data were obtained by Gause (1934) and the curves are simulated, as explained in the text.

under the assumption that the production of the same harmful waste product is the only cause of interaction.

3.2 Relational diagrams

It is sometimes advantageous to summarize the main interrelations of a system in a relational diagram, and to formulate the quantitative aspects at a second stage of actual model building. Such relational diagrams may be presented in various ways, but the conventions introduced by Forrester (1961) prove to be the most convenient in ecology, although they were first developed for the presentation of industrial systems. Forrester assigned special symbols to the various types of variables that may be distinguished in state determined systems. The state variables or the contents of integrals are presented within rectangles, the rates of changes within valve symbols, auxiliary variables within circles and parameters are underlined. The flow of

material is presented by solid arrows and the flow of information by dotted arrows.
The simple system of exponential growth is drawn according to Forrester's conventions in Fig. 4. The amount of organisms is a state

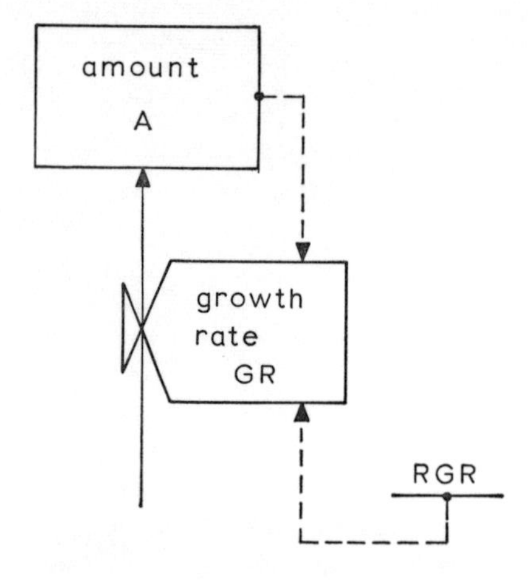

Fig. 4 | A relational diagram of exponential growth, drawn according to the conventions of Forrester.

variable; its value increases by a material flow, whose rate is the growth rate. The dotted line between the state variable and the rate shows that the rate depends (in some way or another) on the state variable and the other dotted line that the rate depends also on a parameter which is here considered a constant.
This figure contains all the interrelations that play a role, but does not consider the details of these. For instance, in the relational diagram, it is still not stated whether the growth rate is proportional to the amount of organisms or to some power of this amount: this is decided upon at a later stage.
The relational diagram for the yeast system is presented in Fig. 5. It is seen that there are three state variables; the amount of the first and second yeast species and the amount of alcohol.
The lines of information flow show directly that the growth of yeast is supposed to depend on the amount of yeast, a relative growth rate and an auxiliary variable: a reduction factor. This reduction factor,

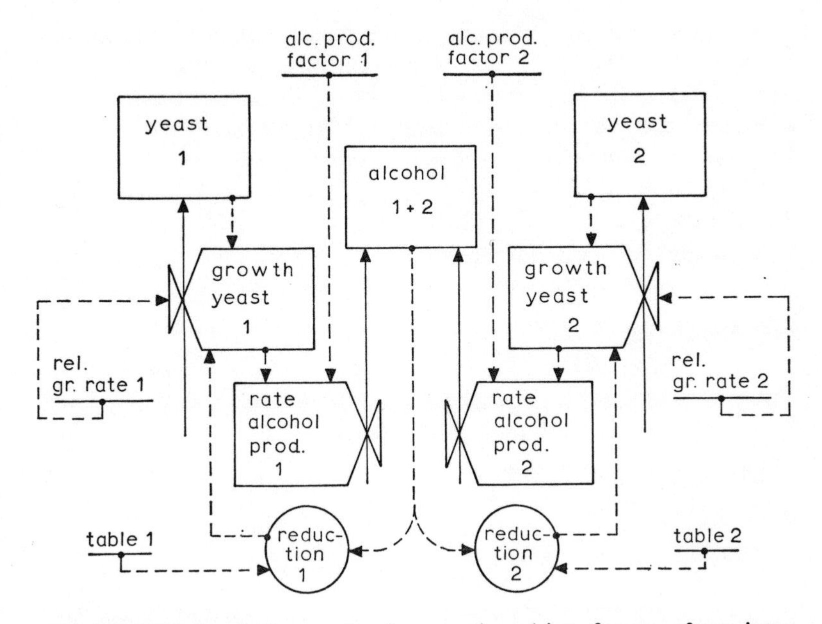

Fig. 5 | A relational diagram for the growth and interference of two interfering yeast species.

in its turn, is given as a function of the amount of alcohol that is present. The relations are, of course, the same for both yeast species although numerical values of parameters and functions may be different. The amount of alcohol increases by the rate of alcohol production of both species. The alcohol production of each species is supposed to depend on the growth rate of the species and on an alcohol production factor.

Exercise 18
In Section 1.4 it is said that rates do not depend on each other in state determined systems. Why is the line of information flow between the rate of growth and the rate of alcohol production not in contradiction with this principle?

Relational models should contain as few details as possible, otherwise they are very difficult to grasp and this defeats their purpose. In studying them, much emphasis should be given to aspects that are not incorporated. For instance, in the present scheme there are no loops that relate the alcohol production directly to the amount of yeast, indicating that the cost of maintenance of yeast cells is not accounted for. The amount of sugar is also not considered because it is assumed to be always available in sufficient amounts.

Exercise 19
Incorporate the aspect of limited food supply in the relational diagram.

3.3 Simulation

The growth of the first yeast species *(Saccharomyces)* is now simulated by stating that the amount of yeast equals

```
Y1 = INTGRL(IY1,RY1)
```
(3.1)

in which

```
INCON IY1 = 0.45
```

is the initial amount of yeast in the arbitrary units, used by Gause, and the rate of yeast growth is given by

```
RY1 = RGR1*Y1*(1.-RED1)
```
(3.2)

The relative growth rate is defined with

```
PARAMETER RGR1 = .......
```

It was observed by Gause that in both species the formation of buds was completely stopped at some maximum alcohol concentration which is given as a percentage by

```
PARAMETER MALC = 1.5
```

The dependence of the reduction factor on the alcohol concentration may now be obtained with an arbitrary function generator

```
RED1 = AFGEN(RED1T,ALC/MALC)
```

The most elementary assumption is that bud formation increases linearly with increasing alcohol concentration, which is introduced with

```
FUNCTION RED1T = (0.,0.),(1.,1.)
```

Exercise 20
Express RED directly in ALC and MALC without using the function generator.

The alcohol concentration itself is the integral of the alcohol production rate which is zero at the initialization of growth:

```
ALC = INTGRL(IALC,ALCP1)                                        (3.3)
INCON IALC = 0.
```

and the alcohol production rate is proportional to the growth rate of yeast:

```
ALCP1 = ALPF1*RY1                                               (3.4)
```

Two values need to be determined now: the relative growth rate and the alcohol production factor. During the early stages of growth, RED is practically zero, so that the growth rate is equal to RGR1 × Y1. This allows a first estimate of RGR1 from the data in Fig. 3 for the monoculture. ALPF1 follows from the observation that growth was terminated when the alcohol concentration equalled 1.5 percent and the amount of yeast about 13 units.

Exercise 21
What is a first estimate of RGR1 in the correct units? What is the value of ALPF1 in the correct units? Is this value only physiologically determined or does it also depend on the volume of water in the vessels with yeast? What is the value of IALC when not only the initial amount of yeast is introduced at initialization, but also the corresponding amount of alcohol? Estimate the same values for *Schizosaccharomyces*, it being known that the alcohol concentration at which the formation of buds is completely inhibited is also 1.5 percent. Which species has the larger alcohol production factor?

The structural equations that describe the growth of the second species (*Schizosaccharomyces*) are, of course, the same as those for the first, so that in a model for concurrent growth it suffices to write them twice: once with a 1 at the end of the relevant symbols and once with a 2; that is except for the equation that describes the alcohol concentration which becomes

```
ALC = INTGRL(IALC,ALCP1 + ALCP2)                    (3.5)
```

This equation holds on the condition that both species interfere only with each other through the production of the same alcohol.
Fig. 6 shows the resulting simulation program with MALC identical for both species and the proper data. In the main program IY1

```
TITLE MIXED CULTURE OF YEAST
      Y1=INTGRL(IY1,RY1)
      Y2=INTGRL(IY2,RY2)
INCON IY1=0.45,IY2=0.45
      RY1=RGR1*Y1*(1.-RED1)
      RY2=RGR2*Y2*(1.-RED2)
PARAMETER RGR1=0.21,RGR2=0.06
      RED1=AFGEN(RED1T,ALC/MALC)
      RED2=AFGEN(RED2T,ALC/MALC)
FUNCTION RED1T=(0.,0.),(1.,1.)
FUNCTION RED2T=(0.,0.),(1.,1.)
PARAMETER MALC=1.5
      ALC=INTGRL(IALC,ALCP1+ALCP2)
      ALCP1=ALPF1*RY1
      ALCP2=ALPF2*RY2
PARAMETER ALPF1=0.12,ALPF2=0.26
INCON IALC=0.
FINISH ALC=LALC
      LALC=0.99*MALC
TIMER FINTIM=150.,OUTDEL=2.
PRTPLT Y1,Y2,ALC
END
STOP
```

Fig. 6 | A simulation program for the growth of two yeast species that interfere through the production of the same waste product (alcohol).

and IY2 are both set to 0.45 units, so that the growth in the mixture is simulated. The two monocultures are simulated in reruns.
FINTIM is set at 150 hours, but the two lines

```
FINISH ALC = LALC
LALC = 0.99 * MALC
```

are inserted to avoid unnecessary 'number-grinding', when the alcohol concentration is close to its maximum. This condition FINISH indicates that the simulation is terminated as soon as the alcohol concentration reaches 99 percent of its maximum value.
The relative growth rates and the alcohol production factors are chosen such that the results of the two experimental monocultures are matched as well as possible. A comparison of the mixtures (Fig. 3) shows that the actual growth of *Schizosaccharomyces* is slightly less than the simulated growth. Barring statistical insignificance, we must

conclude that both species do not interfere with each other's growth through the production of alcohol only, as assumed in the model. It may be that *Saccharomyces* produces some other waste product that is harmful for the other or that *Schizosaccharomyces* produces a waste product that stimulates the other. These possibilities cannot be distinguished from each other without additional information. And as long as this is not available it is a futile exercise to simulate such suppositions.

Exercise 22

Try to reason whether a similar effect could result from the supposition that REDT for the species is not given by

```
FUNCTION RED1T= (0.,0.),(1.,1.)
FUNCTION RED2T= (0.,0.),(1.,1.)
```

but by, for instance:

```
FUNCTION RED1T= (0.,0.),(0.5,0.25),...
  (1.,1.) (Sacch.)
FUNCTION RED2T= (0.,0.),(0.5,0.75),...
  (1.,1.) (Schizos.)
```

If this is too difficult, you may find the answer by simulation.

These simulation programs are conveniently amended. For instance, the yeast cultures may be washed continuously with water that contains sufficient sugar. The integral of the alcohol concentration is then

```
ALC = INTGRL(IALC,ALPF1*RY1 +...
  ALPF2*RY2 - ALC/WSC)
```

in which the washing constant (WSC) is expressed in hours and presents the average residence time of the water in the vials with yeast, as will be shown in Section 6.4.2.

Exercise 23

What is in due course the alcohol concentration and the absolute growth rate of both yeast species for WSC equal to 10 hours?

3.4 Logistic growth

The simulation program in Section 2.2 was developed from the differential equation form. The differential equation form for the present problem will now be derived from the structural equations of the simulation program. This will be done only for situations where the reduction factor is inversely proportional to the alcohol concentration so that (1-RED) may be replaced by (1.-ALC/MALC). Since the alcohol concentration is equal to the amount of yeast times the alcohol production factor according to the Eqns (3.3) and (3.4), it is then possible to rewrite Eqn (3.2) in differential equation form as

$$dY/dT = RGR \times Y \times (1 - Y/YM) \tag{3.6}$$

in which Y is the amount of yeast, T is the time and YM stands for the maximum amount of yeast. This equation may be integrated and this results in

$$Y = \frac{YM}{1 + Ke^{-RGR \times T}} \tag{3.7}$$

Exercise 24

Express YM in MALC and ALCPF. What are the values of YM for both species of yeast? Show by differentiation that Eqn (3.7) is an integrated form of Eqn (3.6). Express the initial amount of yeast in the constant K and YM of Eqn (3.7). Calculate the time course of the growth of *Saccharomyces* and compare the result with the simulated course. Why does the differential equation only hold for situations where the initial amount of yeast is very small, whereas this is not so for the simulation program? (see also Exercise 21).

The growth curve that is described by the differential equation and also presented by the simulated growth curves for the monoculture yeast in Fig. 3 is called the logistic growth curve. This S-shaped curve is symmetrical, but this symmetry hinges on the assumption of inverse proportionality between the reduction factor of growth and the amount of growth that has been made. Especially Lotka (1925) and Volterra (1931) generalized the logistic differential equation for interfering species with the following set of differential equations:

$$dY1/dT = R1 \times Y1 \times (1 - A1 \times Y1 - B1 \times Y2)$$
$$dY2/dT = R2 \times Y2 \times (1 - A2 \times Y1 - B2 \times Y2)$$

In general this set of differential equations cannot be integrated into analytical expressions for Y1 and Y2 as functions of time and therefore it is wiser to leave such simplifying approaches alone and to formulate the problem directly in terms of a simulation model.

Exercise 25

Show to what extent the simulation model for mixed growth of yeast is covered by this set of differential equations. Express the constants R1, R2, A1, A2, B1 and B2 in the constants RGR1, RGR2, ALPF1, ALPF2 and MALC. Which constants of the differential equations are the same? Is this also the case in situations where a species produces a waste product which is only harmful for the other?

3.5 Time constant

The time-interval of integration is adjusted continuously if the integration method of Runge Kutta/Simpson is used. For rectilinear integration, an acceptable time-interval (DELT) may be found by reducing its size to a value where further reduction does not appreciably change the outcome of the simulation.

It is also possible to find a proper time-interval of integration by analysing the simulation program. For this purpose every integral statement and its associated rate is represented by

$$H = INTGRL(\ldots, \pm H/TAU)$$

or in differential form by

$$dH/dT = \pm H/TAU$$

in which the terms of the rate that are independent of H are not considered. TAU, with the dimension time, is the time constant of this particular integration. The time constant of the system at a certain moment is now governed by the integral with the smallest TAU.

It has now been found, that it only makes sense to simulate the dynamic behaviour of a system when a time-interval (DELT) is used, which is about one-tenth of the time constant of the system and that

the value of DELT adjusts to about one-half of the time constant, when the method RKS is used.
If the derivative (H/TAU) has the same sign as the variable (H) itself, there exists a positive feedback. This means that any error is propagated together with an increase of the variable. The feedback is negative in the opposite case and the error will be damped out together with a decrease of the variable itself. The behaviour of the relative error is, however, the same in both situations.
The simplest case of a positive feedback is the exponential growth

$$dH/dT = RGR \times H$$

in which the time constant is equal to the inverse of the relative growth rate. More complicated systems can often be reduced or approximated by a differential equation of the above form. For yeast, the differential equation is

$$dH/dT = RGR \times (1 - RED) \times H$$

In the beginning, RED is small compared to 1, so that the time constant is again 1/RGR. But during growth RED increases, so that the time constant becomes larger. When the method RECT is used, DELT has to be derived from the small value of the time constant in the beginning, but with method RKS its value adjusts during the simulation.
However, it would be wrong to conclude that the time constant of the system approaches infinity when H approaches its maximum value HM. From what has been said about logistic growth, it appears that the equation may also be written as

$$dH/dT = RGR \times H \times (1 - H/HM)$$

when H approaches HM, this is approximately equal to

$$dH/dT = RGR \times HM \times (1 - H/HM) = RGR \times (HM - H)$$

Written for the variable H – HM this is equal to

$$\frac{d(H - HM)}{dT} = -RGR \times (H - HM)$$

This is a differential equation with a negative feedback and again a time constant 1/RGR. This negative feedback governs the time constant of the system if the integral (HM – H) is also considered. The above example is already sufficient to show that it is often difficult, to find the time constant of a system by analysis, unless the system is small. And since its value may be only used for estimating the time-interval of integration, the most practical method is to determine this interval by trial and error.

4 Interference of plants

4.1 Replacement species

The interference of plant species in the field is most conveniently studied by experiments based on the replacement principle.
Experimental plots are divided in small squares. A seed of the first species is placed in each square of one plot and a seed of the second species in each square of another. In this way monocultures of the two species are obtained. On another plot, the seeds of both species are placed alternately in the squares to create a mixture in which half of the space is allotted to one species and the other half to the other. Other mixtures may be obtained by allotting the individual squares to the species in other proportions. The relative seed density of the species, z_1 and z_2 in the mixtures are now defined as the seed density of the species in the mixture divided by its seed density in the monoculture. Obviously, the sum of the relative seed densities z_1+z_2 then equals always 1. The yields of the species in monocultures are represented by the symbols M_1 ($z_1=1$, $z_2=0$) and M_2 ($z_1=0$, $z_2=1$) and the yield of each species in the mixture by O_1 and O_2.

Exercise 26
Show that the replacement principle is not violated when in each square either n seeds of the first species or m seeds of the second are placed.

An experiment is now considered where the individual squares are so large that the two species do not interfere with each other. The seed densities are then low and the yields of both species consequently small. But M_1 and M_2 are of course not necessarily the same. Here the yield of each species in the mixture may be represented by

$$O_1 = \frac{z_1}{z_1+z_2} M_1 \quad \text{and} \quad O_2 = \frac{z_2}{z_1+z_2} M_2 \tag{4.1}$$

The linearity is due to the seeds being so far apart that the plants do not interfere with each other.
The yields may be expressed in dry weight per unit surface or number of seeds per unit surface for seed-forming species. In the latter case the relative reproductive rate of the species may be defined as

$$\alpha_{12} = \frac{O_1/z_1}{O_2/z_2} = M_1/M_2 \tag{4.2}$$

and appears to be equal to the ratio of the yield of the species in monoculture. If α_{12} is 1, the species match each other. If α_{12} is greater than 1, species 1 gains on species 2: the latter eventually disappears from the mixture, if the harvested mixture is resown repeatedly at the original density.
What occurs now if the individual squares on the experimental plots are made smaller and smaller? Then the seed rates of both species increase accordingly and so do the yields. But that is not the only effect. At a certain stage the space allotted to each seed is so small, that the plants interfere with each other. If the species have equal competitive ability one species will not infringe upon the space allotted to the other and Eqn (4.1), resulting in a linear relation between seed rate and yield of the species in the mixture will still be valid. However, in general one species will have more competitive ability and will infringe upon the space allotted to the other. As a consequence, the yield of this species in the mixture will be higher than expected and of the other lower.
Many experiments of this type have been done and the result of one of them with barley and oats is given in Fig. 7. Here the squares were of two sizes: in one experiment 310 cm^2 was allotted to each seed and in the other 31 cm^2. With the wide planting, barley infringed somewhat on the space of oats, but the yield curves were still practically straight. With the narrow planting, however, the yield of barley in the mixtures was relatively high and of oats relatively low, indicating that barley was by far the strongest competitor. The results of this and many other experiments with barley and oats (de Wit, 1960) show that the relative yield total of the mixtures, defined by

$$RYT = O1/M1 + O2/M2 \tag{4.3}$$

is unity.

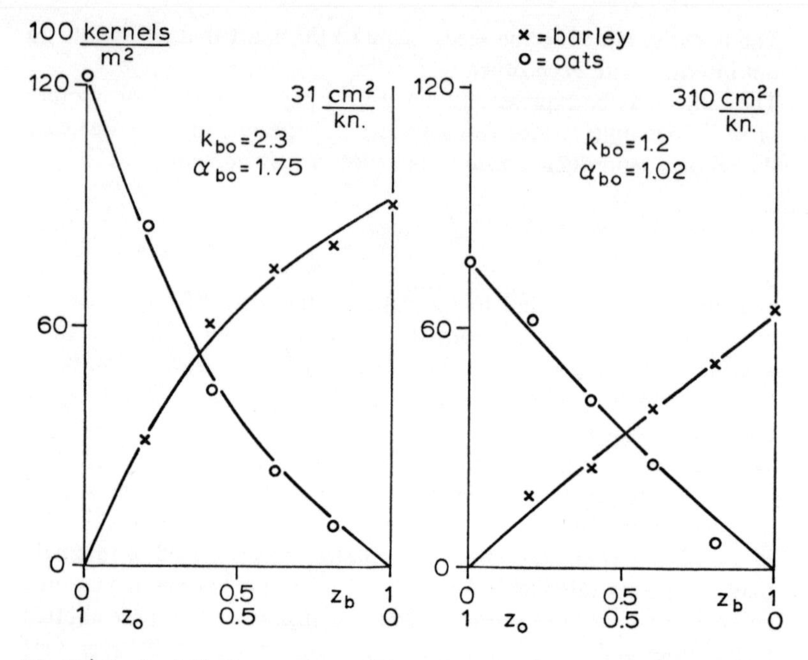

Fig. 7 | Seed yields in number of kernels per m^2 in a replacement experiment of barley and oats at two densities of sowing (de Wit, 1960).

This means that the species appear to be mutually exclusive. This equality may be considered the operational definition of 'competing for the same niche', to use a term out of the field of animal ecology. The relative reproductive rate for seed producing species is now not equal to the ratio of the yields in the monoculture but may be expressed by

$$\alpha_{12} = \frac{O_1/z_1}{O_2/z_2} = k_{12}(M_1/M_2) \tag{4.4}$$

in which k_{12} is the relative crowding coefficient and characterizes to what extent one species infringes upon the space allotted to the other. Eqn (4.3) (with RYT = 1) and Eqn (4.4) may be combined and replaced by

$$O_1 = \frac{k_{12}z_1}{k_{12}z_1+z_2} M_1 \quad \text{and} \quad O_2 = \frac{z_2}{k_{12}z_1+z_2} M_2 \tag{4.5}$$

These equations are similar to the Eqn (4.1), except for the relative crowding coefficient which weights the relative frequency of sowing. Similar relations hold when biomass yields are considered, except that the relative reproductive rate loses its meaning.

Exercise 27

Show that the equations (4.5) are correct by dividing them (O_1/O_2) and by summing the two expressions for O/M. Calculate the relative yield total of barley and oats at relative seed frequencies of 0.2, 0.4, 0.6 and 0.8 from the high density data in Fig. 7. Calculate also the relative reproductive rate and the relative crowding coefficient of barley with respect to oats at the same relative seed frequencies. Which species has the highest yield in monoculture and which species gains in competition?

The yield curves in Fig. 7 have been calculated by assuming that the relative crowding coefficient is independent of the relative seed frequencies and that RYT=1; the agreement between the curves and the experimental data over the whole range of frequencies show that this is a fair assumption. The constancy of the relative crowding coefficient has been confirmed by the analyses of many other experiments (de Wit, 1960; van den Bergh, 1968), so that it is reasonable to state the following. If the relative yield total in replacement experiments equals about 1 over the whole range of seed frequencies, then the relative crowding coefficient may be considered independent of these seed frequencies.

Of course there are also situations where the species do not exclude each other, so that the relative yield total does not equal 1. The equations (4.5) cannot be applied in such situations. For instance, legumeneous species and grass are not mutually exclusive when the first obtains its nitrogen from the air through nitrogen-fixing *Rhizobium* bacteria and the second from the soil and from the first specise (Tow et al., 1966).

The relative yield total (RYT) may be also greater than one when one species has a longer growing period than the other. On the other hand, it has been shown that RYT is smaller than one when one species contains a virus which is harmful to the other (van den Bergh, 1968; Sandfaer, 1970).

4.2 Density of sowing

Replacement experiments between two species and density of sowing experiments of single species have very much in common. This is most conveniently illustrated by considering the results in Fig. 8 of replacement experiments between barley and oats at different pH values of the soil. As far as the relative crowding coefficient is concerned, the two species matched each other at a pH of 4. However, at a pH of 3.7, the relative crowding coefficient of oats with respect to barley was about 2, although the yields of the two species in monoculture were still the same as at the higher pH. Obviously a lower pH affects the competitive ability of barley. This had a detrimental effect on its yield when oats were around to claim the space, but not in monoculture. At a pH of 3.2 the situation was still worse: the relative crowding coefficient of oats with respect to barley increased to 3, whereas the yield of barley decreased to a low level. The physiological cause of the phenomenon is that the root development of barley is much more sensitive to low pH than of oats. The most extreme situation was reached at a still lower pH. Here the relative crowding coefficient of oats with respect to barley increased up to 20, whereas the barley did not grow at all, as reflected by its zero yield in monoculture.

Such a replacement experiment of barley and oats in situations where barley does not grow at all is, in fact, an experiment on the density of

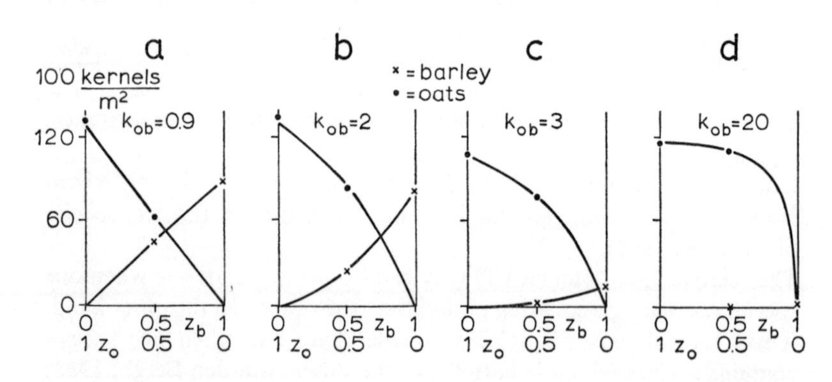

Fig. 8 | Replacement experiments of barley and oats at different pH-KCl values of the soil: 4, 3.7, 3.2, 3.1 for a, b, c, d, respectively (de Wit, 1960).

sowing of oats. In other words density of sowing experiments are a limiting case of replacement experiments. It must be concluded therefore that the equations (4.5) not only describe the results of competition experiments, but those of density of sowing experiments as well. It is only necessary to transform them into a form more suitable for the purpose.

As one species of the replacement series does not grow at all, the second equation may be omitted. The remaining equation is still in an unsuitable form because it is formulated in terms of relative seed frequencies and refers also to the species that is not sown or did not grow at all. A more suitable form is obtained when Z_1/Z_m is substituted for z_1 and $(Z_m - Z_1)/Z_m$ for z_2 in which Z_m is the maximum seed rate used in the experiments and expressed in absolute units, i.e. g m^{-2}. When the subscripts 1 are omitted, the first equation of (4.5) transforms into:

$$O = \frac{B \times Z}{B \times Z + 1} O_m \qquad (4.6)$$

Exercise 28

Derive this formula and express the constants B and O_m in the relative crowding coefficient k and seed rate Z_m and the yield M.

In this equation (4.6) O_m and B are independent of the density of sowing Z. The dimension of Z is number of plants m^{-2} or a similar unit. O_m is the theoretical maximum yield in, for instance g m^{-2}, that is obtained when the seed density is very high and $B \times O_m$ is the yield of a single plant growing alone. B itself has the dimension of m^2 $plant^{-1}$ and may be considered the amount of space that is occupied by a single plant growing alone. The value of O/O_m has a lower limit of 0 and an upper limit of 1.

Exercise 29

Construct a graph from Eqn (4.6) for $O_m = 100$, $B = 0.05$ and Z ranging from 0 to 100. Draw the asymptote O_m and the initial slope $B \times O_m$ of the curve. Mark along the horizontal axis the position where the yield is half of the maximum yield O_m. Mark also the distance $1/B$ along the horizontal axis. Give now expressions for:

$\lim_{Z \to \infty}(O) = \ldots$

$\lim_{Z \to 0}(O/Z) = \ldots$

$\lim_{Z \to 0}(O/O_m) = \ldots$

$\lim_{Z \to \infty}(O/O_m) = \ldots$

The result of a spacing experiment with subterranean clover harvested at various times after planting is given in Fig. 9.
It appears that O_m increases monotonously with time. This is to be expected because the rate of increase of this parameter presents in principle the growth rate of a closed crop surface from the beginning of growth onwards. Under favourable conditions it may be expected that O_m increases with at least 20 g m^{-2} day^{-1}, this being the potential growth rate of most agricultural crops in the Netherlands (de Wit, 1968). The value of B also increases monotonously with time; it represents the (calculated) ability of a single plant to occupy space during its growth and this ability is strongly affected by the stage of

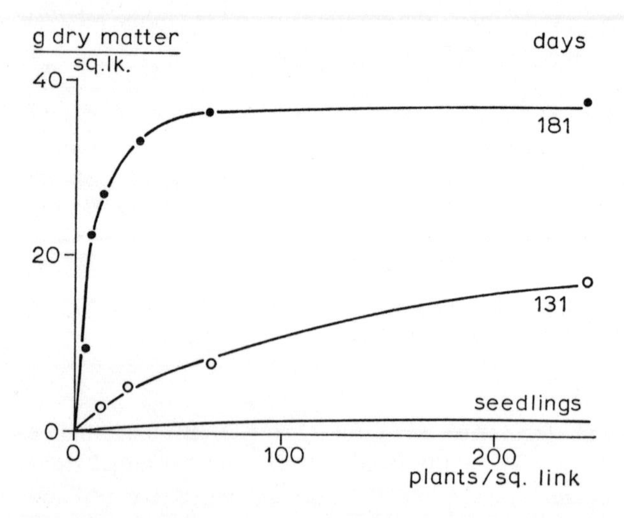

Fig. 9 | A spacing experiment with subterranean clover of Donald (1963), harvested at various times after planting.

development and the distribution of photosynthesis products over the various plant organs.

Baeumer & de Wit (1968) did a spacing experiment with barley and oats on a soil well supplied with nutrients and water. Rows of plants, rather than single plants, were grown at distances of 25 and 100 cm and the dry matter yield was determined at four stages. The results of this experiment are summarized in Table 1.

Table 1 The dry biomass yield in g m^{-2} of barley and oats, sown at 25 and 100 cm on 2 May 1966. Emergence and seedling establishment was completed on 15 May. Field experiment IBS 975, 1966.

Date of harvest	Barley 25	Barley 100	Oats 25	Oats 100
7 June	117	36	81	22
21 June	426	223	319	142
5 July	588	341	503	263
19 July	858	496	789	516

Exercise 30

Calculate the values of B for barley and oats on the four harvesting dates without any assumption regarding O_m.

What is the dimension of B? Draw graphs of O_m and B against time. Which graph has an unexpected form? What are the reasons?

Linearize the curves for O_m, omitting the data points for the first harvesting date and recalculate B for the value of O_m estimated in this way.

The calculated curves of B and O_m against time for barley and oats are given in Fig. 10.

Barley grows somewhat better at low temperatures and its value of B increases during the early part of the growing season more rapidly than for oats. Hence, when both species are grown together, barley occupies relatively more space and by the time oats gets around to

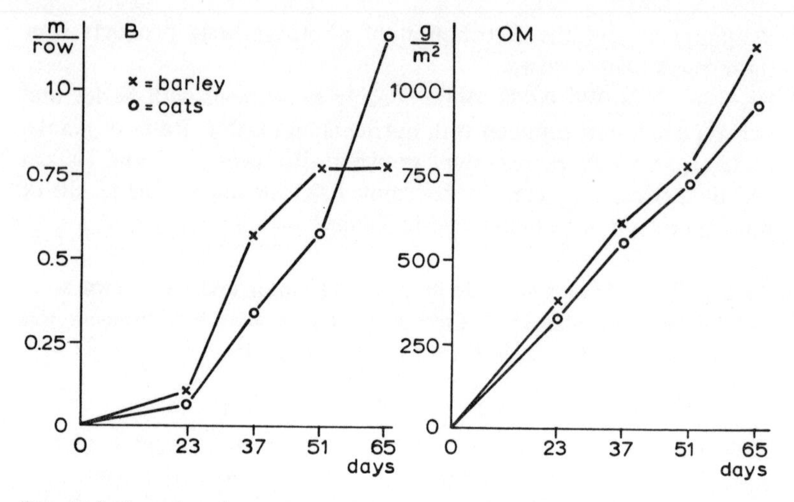

Fig. 10 | Time functions of B and OM for barley and oats, calculated from the data in Table 1.

claim its share, all the space is already occupied. This explains qualitatively why the competitive ability of barley in a mixture with oats is usually higher. Therefore, it may be possible to calculate the mutual interference of both species in a mixture from the course of B and O_m, as determined from density experiments with one species.

4.3 Simulation of plant interference

To arrive at a simulation program for the interference of plants, it is necessary to distinguish the correct state variables and to find expressions for their rate of change. A convenient state variable is the relative space that is occupied by the species, defined as the yield (O) of the species, divided by the maximum yield (O_m) obtained at very high seed density. This relative space is according to Eqn (4.6):

$$RS = \frac{B \times Z}{B \times Z + 1} \tag{4.7}$$

The term relative space is preferred because the term relative yield for this quotient would lead to confusion with the term relative yield

used in the analyses of replacement series. The value of RS ranges from 0 to 1.
The rate of change of the state variable may be found by differentiating RS with time and rearranging the expression. The result is

$$\frac{d(RS)}{dT} = \frac{dB/dT}{B} \times RS \times (1 - RS) \qquad (4.8)$$

This equation is very similar to the equation for logistic growth, derived in Section 3.4; the two main differences being the maximum value of 1 for the state variable and the non-constancy of the 'relative growth rate'.

Exercise 31
Derive the expression for d(RS)/dT. For this purpose, Eqn (4.7) is differentiated, taking into account that B is a variable function and Z a constant function of time. Eqn (4.7) is then used again to eliminate Z. What is the dimension of (dB/dT)/B? Does this relative rate of change increase or decrease with time. What is the expression for B against time when the relative rate of change is constant?

Eqn (4.8) holds for one species. The factor (1 – RS), which may range from practically 1 in the beginning to practically 0 at the end of the growth period, characterizes the reduction of growth under influence of the space that is occupied. When two species are growing together, a situation may be vizualized where plants do not distinguish between occupation of space by one species or the other. Then the relative spaces may be added as to their influence on the growth of each species so that the following set of equations characterize the situation:

$$\frac{d(RS1)}{dT} = \frac{dB1/dT}{B1} \times RS1 \times (1 - SRS)$$

$$\frac{d(RS2)}{dT} = \frac{dB2/dT}{B2} \times RS2 \times (1 - SRS)$$

$$SRS = RS1 + RS2$$

Exercise 32
Construct a relational diagram of this type of plant interference.

The two differential equations are the basis for a simulation program of two species grown in a mixture, which is presented in Fig. 11.

```
TITLE COMPETITION BETWEEN BARLEY AND OATS
INCON DBI1=0.0047,DBI2=0.0033,RSI1=0.002,RSI2=0.002
      RS1=INTGRL(RSI1,(DB1/B1)*RS1*(1.-SRS))
      RS2=INTGRL(RSI2,(DB2/B2)*RS2*(1.-SRS))
      B1=AFGEN(BTB1,TIME)
      B2=AFGEN(BTB2,TIME)
      DB1=DERIV(DBI1,B1)
      DB2=DERIV(DBI2,B2)
      O1=RS1*AFGEN(OMTB1,TIME)
      O2=RS2*AFGEN(OMTB2,TIME)
      SRS=RS1+RS2
PRINT RS1,RS2,SRS,O1,O2
TIMER FINTIM=65.,PRDEL=1.
FUNCTION OMTB1=0.,0.,23.,377.,37.,612.,51.,780.,65.,1132.
FUNCTION OMTB2=0.,0.,23.,333.,37.,552.,51.,724.,65.,956.
FUNCTION BTB1=0.,0.001,23.,0.11,37.,0.574,51.,0.778,65.,0.778
FUNCTION BTB2=0.,0.001,23.,0.076,37.,0.346,51.,0.571,65.,1.17
END
STOP
```

Fig. 11 | A simulation program for interference of two plant species that do not distinguish between the occupation of space by one species or the other.

The function tables for B and O_m are those for the barley(1) and oats(2) experiment of Table 1. DERIV is the only new function that is introduced. This function calculates the derivative of the second argument, here the value of dB/dT from the function of B against time. Like an integral, the derivative has to be initialized and this initial value is given as the first argument of the function.

Exercise 33
Why is it necessary to set the value of B slightly above 0 at emergence? Initialize RSI1, RSI2, DBI1 and DBI2. Is it necessary to initialize the derivative functions accurately?
Compare the results of the simulation graphically with those of the actual competition experiment in Table 2.

Table 2 The dry biomass yield in $g\,m^{-2}$ of barley and oats, sown alternately in rows 25 cm apart. Field experiment IBS 975, 1966.

Date of harvest	Mixture Barley	Oats
7 June	62	30
21 June	235	142
5 July	375	165
19 July	512	308

Inspection of the experimental data in Table 2 shows that barley occupied much more space than oats in the mixture although both species were planted alternately in rows. The simulated results given in the answer to exercise 33 prove that this better performance may be explained by the more favourable course of the B curve for barley during early growth. The higher values of B for oats later appear to be ineffective in the mixture because too much space is already occupied by the barley at the early stages of growth and at this sowing density.

Although this simple model of interference holds for mixtures of some species, it does not always hold in situations where species exclude each other. For instance, in mixtures of short and long peas, it makes a lot of difference to the short peas, whether the space is occupied by other short peas or long peas. In the latter case, practically all light is intercepted by the long neighbours so that the growth of the short peas is almost suppressed. Experimental and simulated results of a competition experiment with these species are given in Fig. 12. To obtain the simulated curves A, it was assumed that the simple model as used in this section for barley and oats was valid. The difference between actual and simulated results is so large that this supposition must be rejected. The curves B were obtained by assuming that the relative space of each species may be weighted according to their respective heights (H1 and H2) which differed at

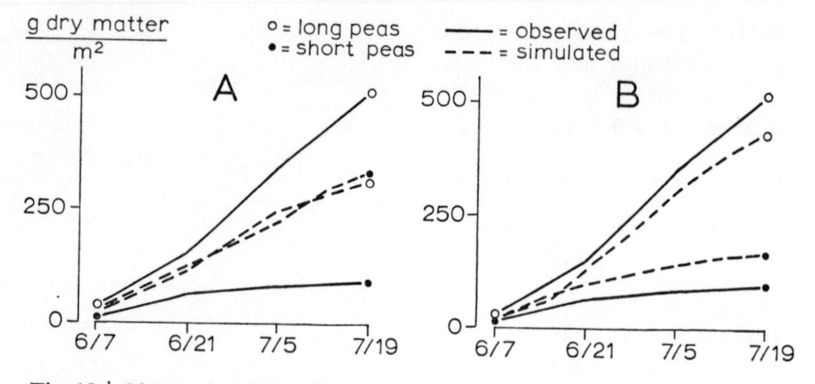

Fig. 12 | Observed and simulated growth of long and short peas in a mixture.
A: without weighting according to height.
B: with weighting according to height.

the end about threefold. This weighting was done according to the equation

$$SRS = 1 - RS1 - (H2/H1) \times RS2$$

for the sum of the relative space. In this way the different light interception is accounted for in a first approximation. The much better agreement shows that this may be a reasonable explanation for the bad performance of the short peas in the mixture.
In situations where species interfere in other ways than by mutual exclusion, it is of course futile to construct a model of competitive interference on basis of data obtained in monoculture only.

4.4 Further modelling aspects

4.4.1 *The* INDEX *and* MACRO *feature*

The simulation programs for yeast growth in the previous chapter and for plant competition in this chapter are given for two species but may be extended to more species. For a mixture of n species, the relevant structural equations must be written n times. Much repeated writing, however, does make the program less clear and often introduces errors, especially if it is necessary to change the structure. Hence any repeated writing that has to be done is best executed by

the computer.
This is done most directly by using the INDEX feature. If for instance, two plant species are competing, the equations for the relative space (RS) are written as

```
RS'1,2' = INTGRL(RSI'1,2',...
  DB'1,2'/B'1,2'*RS'1,2'*(1.-SRS))
```

This is an order for the so-called preprocessor (Section 1.5) to write the equation two times: once with number 1 attached to the symbols, once with number 2. All variables that are different for each species obtain an appropiate number by order of the index '1,2' and all variables, that are the same for the two species, like SRS, remain the same. These have to be defined on their own, in this case by

```
SRS = RS1 + RS2
```

The other equations that describe the growth are given in Fig. 13, which contains a full program for competition between two species. The initial values for the relative space and the slopes of B are defined on parameter cards by

```
INCON  RSI'1,2' = .......,......
INCON  DBI'1,2' = .......,......
```

The four functions must be given separately in FUNCTION tables. The corresponding CSMP program that is written by the preprocessor on the basis of this text, is also given in Fig. 13. Here it can be clearly seen that the INDEX feature is an order for repeated writing or defining of similar texts, parameters or output. In case of more than two species, i.e. four, the statement

```
RS'1,4'=.....
```

generates equations for RS1, RS2, RS3 and RS4.

Exercise 34
Write a program for the growth of 4 yeast species in a mixture, using the INDEX feature.

A similar result may be obtained by using the MACRO-feature. In a MACRO, a part of a process is described in general terms. Every time a MACRO is called upon CSMP writes its full text with the

```
          ***LISTING OF PROGRAM***
     TITLE COMPETITION BETWEEN BARLEY AND OATS
     INCON DBI'1,2'=0.0047,0.0033
     INCON RSI'1,2'=0.002,0.002
           RS'1,2'=INTGRL(RSI'1,2',(DB'1,2'/B'1,2')*RS'1,2'*(1.-SRS))
           B'1,2'=AFGEN(BTB'1,2',TIME)
           DB'1,2'=DERIV(DBI'1,2',B'1,2')
           O'1,2'=AFGEN(OMTB'1,2',TIME)
           SRS=RS1+RS2
     TIMER FINTIM=65.,OUTDEL=1.
     FUNCTION BTB1=0.,0.001,23.,0.11,37.,0.574,51.,0.778,65.,0.778
     FUNCTION BTB2=0.,0.001,23.,0.076,37.,0.346,51.,0.571,65.,1.17
     FUNCTION OMTB1=0.,0.,23.,377.,37.,612.,51.,780.,65.,1132.
     FUNCTION OMTB2=0.,0.,23.,333.,37.,552.,51.,724.,65.,956.
     PRTPLT RS'1,2',SRS,O'1,2'
     END
     STOP
     ENDJOB


              ****CONTINUOUS SYSTEM MODELING PROGRAM****.
       ***PROBLEM INPUT STATEMENTS***
    TITLE COMPETITION BETWEEN BARLEY AND OATS
    INCON DBI1=0.0047,DBI2=0.0033
    INCON RSI1=0.002,RSI2=0.002
    RS1=INTGRL(RSI1,(DB1/B1)*RS1*(1.-SRS))
    RS2=INTGRL(RSI2,(DB2/B2)*RS2*(1.-SRS))
    B1=AFGEN(BTB1,TIME)
    B2=AFGEN(BTB2,TIME)
    DB1=DERIV(DBI1,B1)
    DB2=DERIV(DBI2,B2)
    O1=AFGEN(OMTB1,TIME)
    O2=AFGEN(OMTB2,TIME)
          SRS=RS1+RS2
    TIMER FINTIM=65.,OUTDEL=1.
    FUNCTION BTB1=0.,0.001,23.,0.11,37.,0.574,51.,0.778,65.,0.778
    FUNCTION BTB2=0.,0.001,23.,0.076,37.,0.346,51.,0.571,65.,1.17
    FUNCTION OMTB1=0.,0.,23.,377.,37.,612.,51.,780.,65.,1132.
    FUNCTION OMTB2=0.,0.,23.,333.,37.,552.,51.,724.,65.,956.
    PRTPLT RS1,RS2,SRS,O1,O2
    END
    STOP
OUTPUT VARIABLE SEQUENCE
SRS     B1     DB1     ZZ0002 RS1     B2     DB2     ZZ0004 RS2     O1
O2
OUTPUTS    INPUTS     PARAMS     INTEGS + MEM BLKS    FORTRAN    DATA CDS
15(500)   46(1400)    10(400)      2+   0=  2(300)    12(600)       10
    ENDJOB
```

Fig. 13 | A simulation program for interference of two plant species, written by using the INDEX feature and the CSMP program compiled from this by the preprocessor.

appropriate symbols. A MACRO is therefore not an order to execute a particular computation, but an order to write a part of a simulation program. Just as in a normal simulation program, it is not necessary to present the structural statements in computational order and it may well be that various parts of the MACRO are scattered throughout the computational program after the sorting process. The MACRO for the growth of a plant species may read as follows:

```
MACRO O,RS = GROWTH(BTB,OMTB,DBI,RSI)
  RS = INTGRL(RSI,(DB/B)*RS*(1.-SRS))
  B = AFGEN(BTB,DAY)
  DB = DERIV(DBI,B)
  O = RS*AFGEN(OMTB,DAY)
ENDMAC
```

The first line indicates that there is a MACRO 'GROWTH', in which it is stated how the relative space and the yield (RS and O) depend on functions, variables and initial constants, given or calculated elsewhere in the CSMP program. The ENDMAC line indicates the end of the MACRO. Within the MACRO, the equations of the last section are given but with the numbers 1 and 2 omitted.
The MACRO is invoked by the sentence

```
  O1,RS1 = GROWTH(BTB1,OMTB1,DBI1,RSI1)
```

for species 1 and

```
  O2,RS2 = GROWTH(BTB2,OMTB2,DBI2,RSI2)
```

for species 2.
A program for competition between 2 species and the intermediate CSMP program that is generated are presented in Fig. 14. Detailed comparison of the text shows that three classes of names for variables, parameters and tables can be distinguished. First, those that are mentioned in the statement: these replace the dummy names at corresponding places in the MACRO definition. Secondly those that are used within and outside the MACRO: these remain unchanged and are not necessarily mentioned in the invoking line. Thirdly there are dummy names that are used only within the MACRO: these are replaced by unique names of the type ZZ... in order to avoid double definitions.

Exercise 35
Make a detailed comparison of the 'intermediate' CSMP program written with the MACRO feature, the CSMP program written with the INDEX feature and the original CSMP program for competition between 2 species. It is only in this way that all logical aspects of the MACRO operations can be understood.

```
                ****CONTINUOUS SYSTEM MODELING PROGRAM****
           ***PROBLEM INPUT STATEMENTS***
       TITLE COMPETITION BETWEEN BARLEY AND OATS
       INCON DBI1=0.0047,DBI2=0.0033,RSI1=0.002,RSI2=0.002
       MACRO O,RS=GROWTH(BTB,OMTB,DBI,RSI)
             RS=INTGRL(RSI,(DB/B)*RS*(1.-SRS))
             B=AFGEN(BTB,TIME)
             DB=DERIV(DBI,B)
             O=RS*AFGEN(OMTB,TIME)
       ENDMAC
             O1,RS1=GROWTH(BTB1,OMTB1,DBI1,RSI1)
             O2,RS2=GROWTH(BTB2,OMTB2,DBI2,RSI2)
             SRS=RS1+RS2
       TIMER FINTIM=65.,OUTDEL=1.
       FUNCTION BTB1=0.,0.001,23.,0.11,37.,0.574,51.,0.778,65.,0.778
       FUNCTION BTB2=0.,0.001,23.,0.076,37.,0.346,51.,0.571,65.,1.17
       FUNCTION OMTB1=0.,0.,23.,377.,37.,612.,51.,780.,65.,1132.
       FUNCTION OMTB2=0.,0.,23.,333.,37.,552.,51.,724.,65.,956.
       PRTPLT RS1,RS2,SRS,O1,O2
       END
       STOP
OUTPUT VARIABLE SEQUENCE
SRS    ZZ0001 ZZ0002 ZZ0004 RS1    ZZ0005 ZZ0006 ZZ0008 RS2    O1
O2
 OUTPUTS    INPUTS    PARAMS    INTEGS + MEM BLKS    FORTRAN    DATA CDS
 17(500)   54(1400)   10(400)      2+   0=  2(300)   12(600)        9
       ENDJOB


       TITLE COMPETITION BETWEEN BARLEY AND OATS
       INCON DBI1=0.0047,DBI2=0.0033,RSI1=0.002,RSI2=0.002
             RS1=INTGRL(RSI1,ZZ0004)
             ZZ0004=(ZZ0002/ZZ0001)*RS1*(1.-SRS)
             ZZ0001=AFGEN(BTB1,TIME)
             ZZ0002=DERIV(DBI1,ZZ0001)
             O1=RS1*AFGEN(OMTB1,TIME)
             RS2=INTGRL(RSI2,ZZ0008)
             ZZ0008=(ZZ0006/ZZ0005)*RS1*(1.-SRS)
             ZZ0005=AFGEN(BTB2,TIME)
             ZZ0006=DERIV(DBI2,ZZ0005)
             O2=RS2*AFGEN(OMTB2,TIME)
             SRS=RS1+RS2
       TIMER FINTIM=65.,OUTDEL=1.
       FUNCTION BTB1=0.,0.001,23.,0.11,37.,0.574,51.,0.778,65.,0.778
       FUNCTION BTB2=0.,0.001,23.,0.076,37.,0.346,51.,0.571,65.,1.17
       FUNCTION OMTB1=0.,0.,23.,377.,37.,612.,51.,780.,65.,1132.
       FUNCTION OMTB2=0.,0.,23.,333.,37.,552.,51.,724.,65.,956.
       PRTPLT RS1,RS2,SRS,O1,O2
       END
       STOP
```

Fig. 14 | A simulation program for the interference of two plant species written by using the MACRO feature. The intermediate CSMP program, produced by the CSMP compiler is also given. The text of this intermediate program is not printed by the computer.

Exercise 36
Only for readers that are familiar with FORTRAN.
What are the principle differences between a MACRO and a SUBROUTINE?

One may wonder why two methods are being developed to make simulation programs more lucid and to avoid repeated writing of structural statements, and the more so because these methods seem very much alike. However, this similarity is only so in the context of the present small programs which are written for illustrative purposes. Later it will become evident that each method has its own field of use.

4.4.2 *The* INITIAL *and* DYNAMIC *section*

The initial values for the relative space (RS) and the derivative of the space occupied by a single growing plant (DB) must be calculated before the simulation models discussed in the previous sections can be applied. In order to avoid errors and again to promote the clarity of the simulation, it is advantageous to incorporate this computation in the simulation program. This can be done again most conveniently by a MACRO, in which it is defined how RSI and DBI depend on the distance of sowing (DIST) and the function for B (BTB):

```
MACRO   RSI,DBI = BEGIN(BTB,DIST)
        RSI = BI/DIST
        BI = AFGEN(BTB,0.)
        DBI = (AFGEN(BTB,1.)-BI)/1.
ENDMAC
```

Exercise 37
Show that RSI can be set equal to BI/DIST and that the expression for DBI is correct.

The computational procedure, contained in this MACRO, has to be done only once for each species before the actual simulation is started. This is done by distinguishing an initial section of the simulation model which starts with an INITIAL card and ends with a DYNAMIC card.

The above MACRO is invoked two times within this INITIAL section.
After this initial section, the normal dynamic structural statements of the simulation program are entered.

Exercise 38
Write a simulation program for the growth of three barley cultivars, assuming that B for the second variety and O_m for the third variety increase half as fast with time as for the first variety. Assume that the species are sown in rows 40 centimetres apart in a 1:1:1 ratio. Be careful about the value of DIST. Make use of the MACROs and the INITIAL section. Write the same program with the INDEX feature.

---

5 Growth and competition of Paramecium

5.1 Description of the system

Paramecia are protozoa: unicellular organisms that live in water and feed on bacteria. Gause (1934) did a series of experiments with the species *P. aurelia* and *P. caudatum* in monoculture and in mixture to study the principles of their mutual interference.
The species were grown in test tubes with 5 cm^3 of Oosterhout's balanced physiological solution, buffered at pH 8.0. The medium was changed daily by centrifuging to separate the protozoa from the liquid with the waste products and the remaining food. A standardized amount of bacteria was added in the new solution as daily food. Just before centrifuging the solution was carefully stirred and one tenth of the volume of liquid was taken out in which the number of protozoa were counted. Hence at the beginning of each day the number of protozoa was about nine-tenth of the number at the end of the day before.

Exercise 39
Why not exactly nine-tenth?

Two series of experiments were done, in the one loop experiment one standardized loop of bacteria was given each day and in the half-loop experiment a half of the standardized loop of bacteria was given. In both series, the species were grown in monoculture and in mixture. The monocultures were started with 20 protozoa of the species concerned and the mixed culture with 20 protozoa of each species. The number of protozoa counted in the sample throughout a period of 16 days are given in Table 3.

Exercise 40
Plot the results on graphs and save these for a first estimation of parameters, later on.

Tablə 3 Numbers as sampled by Gause

Day of the experiment	Monoculture				Mixed culture			
	P.aurelia 0.5 cm^3		P.caudatum 0.5 cm^3		P.aurelia 0.5 cm^3		P.caudatum 0.5 cm^3	
	one loop	half loop	one loop	half loop	one loop	half loop	one loop	half loop
0	2	2	2	2	2	2	2	2
1	6	3	6	5	10	4	5	8
2	24	29	31	22	29	29	15	20
3	75	92	46	16	68	66	32	25
4	182	173	76	39	144	141	52	24
5	264	210	115	52	164	162	40	—
6	318	210	118	54	168	219	32	—
7	373	240	140	47	248	153	36	—
8	396	—	125	50	240	162	40	21
9	443	—	137	76	—	150	32	15
10	454	240	162	69	281	175	20	12
11	420	219	124	51	—	260	30	9
12	438	255	135	57	300	276	12	12
13	492	252	133	70	—	285	16	6
14	468	270	110	53	—	225	20	9
15	400	240	113	59	260	222	12	3
16	472	249	127	57	294	220	9	0

The number of protozoa in the monoculture reached a maximum and stayed there, just as for yeast. The growth of yeast ceased because of the accumulation of waste products. But this cannot be the cause of stabilization in this case, since the waste products were removed every day by centrifuging. It stands to reason that here the ultimate size of the population was limited by the daily food supply. In the equilibrium situation this supply was then just sufficient to maintain the population and to replace the ten percent that was removed by sampling. In the mixed culture one of the species vanished, whereas the other survived at the same level as in monoculture. This competitive phenomenon has to be understood by a further analysis of the system.

To arrive at a quantitative description of the relevant growth and death processes, some assumptions have to be made. First it is assumed that a fixed ratio exists between the number of newly grown protozoa and the amount of food that is consumed. This ratio is called the conversion factor of food (CONVF) and has the dimension of number of protozoa per loop of bacteria. Second, it is assumed that there is a natural death rate which is proportional to the number of protozoa, so that it can be characterized by a constant relative death rate (RDR), which is independent of the density. The rate of food consumption (CNRT) is assumed to be proportional to the number of protozoa (H), the density of food (FOOD) in the medium and the rate at which the protozoa search the water for food (RSW). The density of food is the amount of food (AFOOD) divided by the volume. However, this rate per protozoa cannot exceed the maximum digestion rate of food (MRDIG), because the protozoa may meet food in excess of their rate of intake and digestion.

Exercise 41

Determine the dimensions of the mentioned state, rate, and auxiliary variables and parameters and classify these according to type. Use as basic units: day, loop, protozoon, volume of test tube.

Construct a relational diagram for the growth of one protozoa species, taking into account that each day the population is sampled and the food is renewed. Show that the assumption of a constant relative death rate is mathematically equivalent to the assumption that food is needed to maintain the protozoa.

5.2 A simulation program

As done previously for the competition between plants, the dynamics of one species will be described in a `MACRO`, which is then invoked for each species with the appropriate names. The output variables of the `MACRO` are the number of protozoa `(H)`, the rate of food consumption `(CNRT)` and the size of the sample `(SPLE)`. The input variables are the rate of searching the water `(RSW)`, the conversion

factor of food (CONVF), the maximum digestion rate (MRDIG), the relative death rate (RDR) and the initial size of the population (HI). The moment of feeding and sampling (FDTIME) and the density of food (FOOD) are the same for both species, so that these are defined in structural statements outside the MACRO and do not appear in the MACRO definition.
The MACRO is as follows:

```
MACRO H,CNRT,SPLE=GROWTH(RSW,CONVF,...
  MRDIG,RDR,HI)
```

The amount of protozoa is now defined by

```
H=INTGRL(HI,AGR)
```

The actual growth rate (AGR) is the difference between the net growth rate (NGR) and the rate of sampling (RSAM):

```
AGR=NGR-RSAM
```

and the net growth rate (NGR) is the difference between the gross growth rate minus the natural death rate (DR):

```
NGR=CNRT*CONVF-DR
DR=RDR*H
```

In calculating, the consumption rate of food (CNRT), the maximum digestion rate must be accounted for. This is done by an AMIN1 function, which takes the minimum of its arguments:

```
CNRT=H*AMIN1(MRDIG,RSW*FOOD)
```

Exercise 42
Draw a graph of the consumption rate of food (CNRT) against the density of food (FOOD) for arbitrary values of MRDIG, RSW and H. How does this graph change with changing MRDIG, H or RSW. For which value of FOOD does CNRT equal zero and which value of FOOD does NGR equal zero? Reason why this expression does not contain the amount of protozoa (H).

The calculation of the rate of sampling (RSAM) raises some problems because it is a discontinuous process. The sampling occurs only once a day and is zero for the rest of the time. The sample size is defined with

```
SPLE=FDTIME*0.1*H
```

in which `FDTIME`, as defined outside the `MACRO`, is one during one time-step at the end of the day and otherwise zero. To let the sampled quantity vanish during one time-step, the rate of sampling must be defined as the size of the sample divided by the time-step `DELT`:

```
RSAM= SPLE/DELT
```

as is seen from calculating

$$X_{t+\Delta t} = X_t - (0.1X_t/\Delta t)\Delta t$$

The `MACRO` is now terminated with

```
ENDMAC
```

In the main program, the `MACRO` is called for twice: once for the species *P. aurelia*

```
HA,CNRTA,SPLEA=GROWTH(RSWA,CONVFA,...
  MRDIGA,RDRA,HIA)
```

and once for the species *P. caudatum*

```
HC,CNRTC,SPLEC=GROWTH(RSWC,CONVFC,...
  MRDIGC,RDRC,HIC)
```

In the main program FDTIME is defined by

```
FDTIME=IMPULS(1.,1.)
```

This function has the value 1 at the moment indicated by the first argument and subsequently at intervals defined by the second argument. The rest of the time, the function equals zero. The variable `FDTIME` is used within the macros to define the moments of sampling and outside the `MACRO` also to replenish the food at daily intervals, according to

```
PARAMETER VOLUME=1
FEED=FDTIME*(L-AFOOD)/DELT
AFOOD=INTGRL(L,FEED-CNRTA-CNRTC)
FOOD = AFOOD/VOLUME
```

L is the amount of food given daily after removal of the food that is left over from the previous day and either equal to 1 or 0.5 loop of bacteria. The amount of food during the day is continuously diminished by consumption by the *P. aurelia* and *P. caudatum* species, but only once a day replenished to the original level.

Exercise 43
Why is VOLUME equal to 1 rather than 5 cm^3? Why is it advisable to distinguish between AFOOD and FOOD?

Due to the discontinuity in the food supply and in the sampling it is necessary to integrate according to the

```
METHOD RECT
```

and to specify DELT also on the TIMER card:

```
TIMER FINTIM=16, DELT=0.01, OUTDEL=1
```

Exercise 44
Why is it impossible to use integration methods that adapt the size of the time-interval DELT to the rate of changes of the integrals? (See also Section 2.3).

For comparison with Gause's data it suffices to print the size of the samples SPLEA and SPLEC each day, but a more frequent printing of population numbers is necessary to study the behaviour of thc simulated populations during the day. To complete the program all initial values and parameters must be defined on parameter cards. There are eight parameters: CONVF, RSW, RDR and MRDIG that have to be derived from the experimental data and must be substituted in the simulation program. In principle, these can be found by trial and error, using some goodness of fit criterion to the observational data. But such a procedure can be started only in practice when the order of magnitude of all the variables concerned are known from a preliminary analysis of the data.

Exercise 45
Why?

5.3 First estimation of parameters

Gause observed that at first the medium remained opaque during the whole day, but that later the medium became transparent within a few hours after the addition of new food. From this he concluded that all food was consumed rapidly, once the size of the population was not far from its maximum. Obviously there is sufficient time for digestion and searching and the maximum size of the population does not depend on the rate of digestion of the food or on the rate of searching water. Instead it depends only on the amount of food given, the conversion factor for food, the relative death rate and the rate of sampling. About $H \times (RDR + 0.1)$ number of protozoa die or are sampled and $CONVF \times L$ number grow in the monoculture in a day when the daily food is consumed completely.
In the equilibrium situation, these quantities are equal so that $CONVF \times L = H_{eq} \times (RDR + 0.1)$
This equation contains two unknowns; CONVF and RDR, so that another equation is necessary to estimate their values. This second equation can be obtained by considering the growth rate (GR) at the moment that three-quarters of the maximum population size is reached. This growth rate may be estimated from the experimental data and is equal to:

$$GR = CONVF \times L - 0.75 \times H_{eq} \times (RDR + 0.1)$$

This because it was observed by Gause, that the food was exhausted well within a day at this density.
Combining both equations allows a first estimate of CONVF and RDR.
The rate of searching the water (RSW) and the maximum rate of digestion (MRDIG) are estimated from the dynamics of the populations at the beginning of the experiment. During the early stages, the number of protozoa is so small that the concentration of bacteria stays practically the same during the whole day. It may be seen now from the data that the initial growth rates of the 0.5 and 1 loop series with *P. aurelia* are about the same and this means that the maximum digestion rate is at least reached at the 0.5 loop concentration. In other words, at this level

$$MRDIGA = 0.5 \times RSWA$$

but instead of 0.5, a lower value could be more appropriate.
This is again an equation with two unknowns, so that another equation is necessary to make a first estimate of both parameters. This second equation can be obtained by considering the initial relative growth rate of the one loop series. This relative growth rate can either be read from the data or it can be set equal to

$$RGR = MRDIG \times CONVF - RDR - 0.1$$

at least as long as the bacterial concentration is so high that the maximum digestion rate is maintained during the day.
For *P.caudatum* the relative growth rate of the 1 loop series is higher than of the 0.5 loop series, so that there is no certainty that the maximum rate of digestion is reached at a bacterial concentration of 1 loop per volume. A first estimate of the parameters may be obtained here by assuming that

$$MRDIGC = 1 \times RSWC$$

but instead of 1, a higher value could be more appropriate.

Apart from the uncertainty about the exact value of the constant in the equation for the maximum digestion rate, the estimation procedure is also unfavourably affected by the large scattering of the data. This makes it difficult to arrive at a value for the initial relative growth rate. It is therefore still worthwhile to inspect the system for other interrelations between the constants.
These are obtained from the observation that the maximum number of *P. aurelia* in both the one loop and 0.5 loop series is about 4 times higher than the number of *P. caudatum* so that (probably) the *P. aurelia* individuals are about 4 times smaller. Thus, it is logical to assume at first that the conversion factor of food, as recalled with the unit protozoon $loop^{-1}$, is 4 times larger for *P. aurelia* and that the maximum rate of digestion, in the unit loop $protozoon^{-1}$ day^{-1} is 4 times smaller.

Exercise 46

Take the graphs that were drawn for the monocultures in Exercise 40 and estimate for both species and both series, the maximum population size (H_{eq}), the growth rate (GR) at the moment that the population equals 0.75 of the maximum and the initial relative growth rate (RGR). Calculate the parameters CONVF, RDR, MRDIG and RSW for both species and both series independently with the 2×2 equations given. Make first estimates of these parameters for both species, taking the size of the individual protozoon into account. Try to find as many reasons why these first estimates may be considerably in error.

5.4 Final adaptation of parameters

There are many reasons why the first estimates, especially of the rate of searching and the maximum rate of digestion are very rough indeed. It is therefore necessary to improve on these by comparing the results of simulation runs with the actual results. In principle the results of the monocultures should only be used for this purpose, but it appears that the scattering of the observational data is so large that it is very difficult to arrive at sufficiently accurate estimates for the parameters. Fortunately, the results of the competition experiments are also available to improve the estimates. When these results are used, it should be realized that it is then implicitly stated that the interference between both species as proposed in the model is correct, so that a comparison between simulated and actual results of the competition series cannot be used to validate this assumption. However, the large scattering of the observational data necessitates this way of working.
Further simulations show that the course of *P. caudatum* in the mixture as characterized by the time at which the maximum population size is reached and the rate of its decline in later stages, is especially governed by the ratio between the searching rates of the water by both species and by the ratio of the relative death rate and the conversion factors. In other words, the differences between both species in this respect are especially manifest in the competitive situation.

Exercise 47
Explain why this is so.
Finalize also the simulation program and try to find better estimates of the parameters by trial and error.

After a considerable amount of experimentation with the simulation program it appears that the best agreement between simulated and actual results, as judged visually on graphs, is obtained with the parameter values listed in Table 4.

Table 4 Parameter values for *P. aurelia* (A) and *P. caudatum* (C)

	A	C	
Relative death rate (RDR)	0.45	0.45	day^{-1}
Conversion factor (CONVF)	3000	750	prot. $loop^{-1}$
Saturation level (MRDIG)	$.56\ 10^{-3}$	$2.25\ 10^{-3}$	loop.$prot.^{-1}$ day^{-1}
Rate of searching water (RSW)	.006	.006	volume.$prot.^{-1}$ day^{-1}

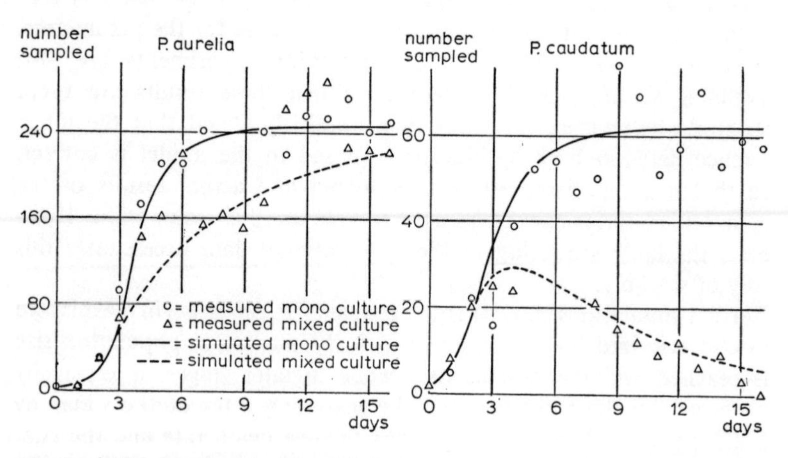

Fig. 15 | Simulated and observational results for the one loop experiment with *P. aurelia* and *P.caudatum.*

The simulated and observational data for the one loop series are summarized in Fig. 15, to show that within the limits of accuracy governed by the scattering of the data there is a good agreement and that the results can at least be understood by assuming that the species only affect each other by competing for the same food. The largest species, *P. caudatum*, loses in competition probably because the rate of searching the water does not increase proportionally with the size of the protozoa, so that per unit biomass of protozoa less bacteria are available for the larger animal. This is likely, because the surface weight ratio, and with this the mobility and the chance of meeting bacteria, is considerably reduced. This competitive advantage is obvious in situations where the concentration of bacteria is small. At higher concentrations, the consumption is governed by the maximum rate of digestion which is four times larger for the four times larger species, so that the species match each other in this respect.

In the analysis of the original experiments of Gause, it was taken into account that during the early stages the concentration of protozoa was so small that the food level hardly decreased during the day and that during later stages the food was rapidly depleted. Although not observed by Gause, we are now in a position to consider in more detail the daily course of food concentration and number of protozoa because these have been simulated. Some of these simulated results are presented in Fig. 16 for further inspection.

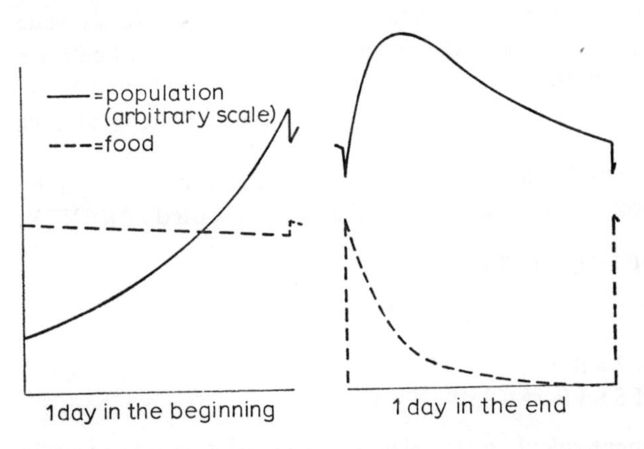

Fig. 16 | Simulated course of growth and food supply of a *Paramecium* species in monoculture during a day at the beginning and the end of the experiment.

Exercise 48

Why is the growth of protozoa during the beginning of the experiments nearly exponential? Why, at the end of the experiment, is the maximum size of the population at some time during the day, larger than the population size, observed at the end of the day? Which is higher, death through natural causes or through sampling?

5.5 Stochastic aspects

The simulation program presented in the previous sections is fully deterministic and does not explain at all the large scatter of the observational data. There are, however, two stochastic phenomena that are accessible for further analysis. These are the sampling process and the death process.

As far as the sampling process is concerned, it was assumed that exactly 1/10 of the population is taken away when 1/10 of the solution is removed. However, this is not true. The protozoa are, after stirring, randomly distributed throughout the solution so that either more or fewer protozoa than the average may actually be found. To simulate the actual number that are in the sample, this number must be drawn out of a probability function around the average. Since the number of protozoa may be small, the probability function of Poisson may be used.

This function can be introduced into the simulation program by replacing the statement for the sample size in the `MACRO GROWTH`:

```
SPLE=FDTIME*0.1*H
```

by the statements

```
AVSMP=0.1*H
SPLE=POISS(P,AVSMP,1.)
```

The first statement calculates the size of the average sample at every time-interval and the second statement is a function that assigns an appropriate random number to the sample size. The value of the first variable in this argument is an odd number, to be specified on a parameter card (outside the `MACRO`) and is necessary to start the process of generating random numbers. The second variable in the argument is the average number of protozoa in the sample and the number 1 indicates that the sample is taken with an interval of one day.

As far as the death process is concerned, the amount of protozoa that die during one time step (AD) is on the average:

```
AD=H*RDR*DELT
```

and the random number that dies is accordingly

```
RD=POISS(P,AD,DELT)
```

The third variable in the argument is DELT because death occurs every time-step. The rate of dying is now calculated from the amount that dies by dividing again by DELT with

```
DR=RD/DELT
```

These three statements replace the statement

```
DR=RDR*H
```

in the original MACRO GROWTH.

Results of some simulations are presented in Fig. 17, which shows

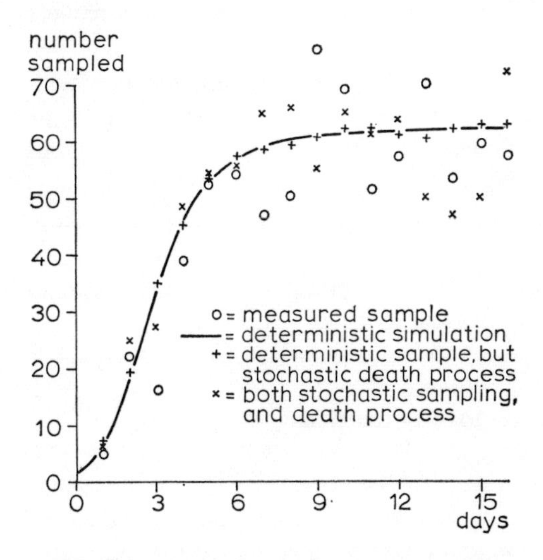

Fig. 17 | Observational and simulated results of *P. caudatum* in the one loop experiment under various assumptions regarding the operation of random processes.

the growth of *P. caudatum* in the one loop series. The solid line is the growth curve obtained by deterministic simulation. The Roman crosses are the simulated results with a deterministic sampling process and a random death process and the Greek crosses present the simulated results that are obtained with a stochastic sampling and stochastic death process. The open dots are the observed data. It must be concluded from these results that the main contribution to the variability is due to the method of sampling and that the quality of the experiment would have been very much improved if some method of measuring the whole population had been introduced.
The scattering due to the stochastic sampling is much larger than due to the stochastic death process, although about 45 percent of the population dies during one day and only 10 percent is sampled. A simple calculation can explain this. Let the equilibrium population be 1000 individuals. The average sample size is then 100 and the standard deviation is according to the binomial probability function $\sqrt{0.9 \times 0.1 \times 1000} = 9.5$. Each day an average of 450 individuals dies out of 1000 and the standard deviation of this number is $\sqrt{0.45 \times 0.55 \times 1000} = 16$. Because one tenth is sampled this reduces to a standard deviation of 1.6 in the sample and this is only one sixth of the standard deviation caused by sampling process itself. Moreover, the death process is distributed over the day so that some deviation may be even levelled by negative feedback throughout the day.

Exercise 49
Explain now why the scattering of the observational data for *P. aurelia* is much smaller than for *P. caudatum*.

5.6 The programming of probability functions

To simulate stochastic processes, CSMP contains a so-called random generator that generates numbers between 0 and 1 out of a standard uniform probability function and a Gaussian generator that generates numbers out of a normal probability function with a specified average and standard deviation. The language does, however, not contain a Poisson generator, so that such a generator has to be introduced by

the user. This is most conveniently done in the form of a MACRO and in this section the content of this MACRO is described. Unfortunately this has to be done in a way that is only understandable for the reader who has some knowledge of FORTRAN and probability calculus.
The heading is:

```
MACRO N=POISS(P,MEAN,PERIOD)
```

DO loops and IF statements as such cannot be sorted by CSMP, so that the statements are given in computational order. This is indicated by the card:

```
PROCEDURAL
```

If the time is not equal to n times PERIOD, the sampling need not be executed and N equals zero:

```
N=0
IF(IMPULS(0.,PERIOD).LT.0.5) GO TO 1
```

whereby 1 is a CONTINUE statement at the end of the MACRO. In case the expectation value is larger than 25, the Poisson distribution is sufficiently approximated by a Gauss distribution with a standard deviation equal to the square root of the average.

```
IF(MEAN.LT.25.) GO TO 2
N=GAUSS(P,MEAN,SQRT(MEAN))
GO TO 1
2 CONTINUE
```

The Gauss function is a CSMP function that executes the random choice out of a normal distribution. P can be any odd integer. The second and the third argument represent the average and the standard deviation, respectively.
Below a number of 25 the deviation between the Poisson distribution and the Gauss distribution becomes too large. To execute the selection from the Poisson distribution a number is first drawn between 0 and 1 according to standard uniform probability function. This is done by a CSMP function:

```
LOT=RNDGEN(P)
```

P is again the odd integer.
Then this number is used to read the output from a cumulative

Poisson distribution function. The cumulative Poisson distribution is obtained by a series development. The probability of a number to be smaller than or equal to 0, 1, 2, 3... is given by $e^{-z}(1+z/1!+z^2/2!+z^3/3!...)$ where z is its average.
This is programmed as follows:

```
SUM=1.
PROD=1.
EMINZ=EXP(-MEAN)
DO 4 J=1,100
IF(LOT.GT.SUM*EMINZ) GO TO 3
N=J-1
GO TO 1
3 CONTINUE
PROD=PROD*MEAN/J
SUM=SUM+PROD
4 CONTINUE
```

Then the MACRO is concluded by

```
1 CONTINUE
ENDMAC
```

The PROCEDURAL card also ensures that all statements of the MACRO are sorted as one block at a place where P, MEAN and PERIOD are available and N is needed, as indicated by the MACRO definition card.

By using a Poisson probability distribution function which is for higher numbers replaced by the Gaussian function, numbers higher than the total number of individuals in the population may be drawn. Chances that this occurs are very small when the death process is considered, but may not be negligible in the sampling process. This problem does not exist when the sampling process is formulated on basic principles.

For this purpose, the protozoa in the solution are considered analogous to the black balls and the volumes of water equal to the volume of protozoa analogous to the white balls in the traditional jar with coloured balls. The following symbols can now be defined:

N: the total number of volume elements and n: the number drawn,

B: the total number of protozoa and b: the number drawn (black balls),

W: the total number of volume elements water and w: the number drawn (white balls).

Hence, $N = B + W$ and $n = b + w$.

According to basis theory, the number of combinations of drawing a number of n balls out of a total of N is:

$$\frac{N!}{(N-n)!\,n!} \tag{5.1}$$

Similar expressions hold for the white and black balls, so that the number of combinations of drawing b black balls and w white balls equals the product

$$\frac{B!}{(B-b)!\,b!} \times \frac{W!}{(W-w)!\,w!} \tag{5.2}$$

To obtain the probability of obtaining b black and w white balls in the sample, this expression must be divided by the total number of combinations. This gives

$$\frac{B!}{(B-b)!\,b!} \times \frac{W!}{(W-w)!\,w!} \times \frac{(N-n)!\,n!}{N!} \tag{5.3}$$

In the present situation, the volume of water is infinite with respect to the volume of paramecia, so that W and w are infinite with respect to B and b.

Hence when a fraction f of the volume is sampled the total number of volume elements (water and protozoa) is fixed according to

$$n = f.N$$

Since also $W = N - B$ and $w = n - b$, expression (5.3) for the probability can be transformed into

$$\frac{B!\,(N-B)!\,(f\cdot N)!\,((1-f)\cdot N)!}{(B-b)!\,b!\,((1-f)\cdot N - B + b)!\,(f\cdot N - b)!\,N!} \tag{5.4}$$

which approaches to

$$\frac{B!\,f^{b}(1-f)^{B-b}}{(B-b)!\,b!} \tag{5.5}$$

with increasing N.

This is a binomial probability distribution function.
The chance to find 0, 1, 2,... paramecia in the sampled volume is now

Number	Chance
0	$(1-f)^B$
1	$B\,f\,(1-f)^{B-1}$
2	$B\,(B-1)\,f^2\,(1-f)^{B-2}/2$

The sampling may now be programmed as follows:

```
MACRO N=BINOM(B,F,P,PERIOD)
```

N is the number which is actually drawn, B is the total number of paramecia in the vessel, F is the fraction of the liquid which is taken out, P is some odd integer and PERIOD is the interval of sampling.

```
PROCEDURAL
N=0.
IF(IMPULS(0.,PERIOD).LT.0.5) GO TO 100
LOT=RNDGEN(P)
PROD=(1.-F)**B
SUM=0.
DO 400 J=1, 100
SUM=SUM+PROD
PROD=PROD*(B-J+1)*F/(J*(1.-F))
IF(LOT.GT.SUM) GO TO 400
N=J-1
GO TO 100
400 CONTINUE
100 CONTINUE
ENDMAC
```

Still one remark should be made. If the expectation value of the sample is small and f is small, the expression for the probability distribution may be simplified even more. The expectation value is then $f \times B$. If this product stays at a constant low value, then f decreases with increasing B. The ratio $B!/(B-b)!$ approaches then B^b and the power $(1-f)^{B-b}$ approaches $(1-f)^B$ which can be replaced by $e^{-f \times B}$. Substitution in the expression for the binomial distribution Eqn (5.5) gives

$$\frac{(f \times B)^b \times e^{-f \times B}}{b!} \tag{5.6}$$

Replacing the expectation value $f \times B$ by z gives

$$\frac{z^b e^{-z}}{b!} \tag{5.7}$$

which is the Poisson probability distribution function.

6 Modelling of development, dispersion and diffusion

6.1 Introduction

In Chapter 1 it was stated that systems ecology is based on the assumption that the state of an ecosystem at any particular time can be expressed quantitatively and that changes in the system can be described in mathematical terms. Various models of ecosystems were given and in all examples it was possible to use a very limited number of state variables and associated rate equations. This is not surprising. Yeast and paramecium are simple organisms and the responses as a population are hardly dependent on such attributes as size and stage of development. The small grain example concerns more complicated organisms that are synchronized in time and whose responses strongly depend on size, stage of development and physiological conditions and on the continuously changing physical environment. However, in this case the problem was simplified by a model with a limited number of state variables by only treating the interference of similar plant species. No attempt was made to construct a predictive model of the growth and development of form and function of the single species.
Although we may accept that the ultimate purpose of biology in general and ecosystems analysis in particular, is the construction of models that predict growth and development of single and interfering species in natural conditions, we must admit that at present this goal is unrealistic. The knowledge of the relevant processes is quantitatively, but also qualitatively far too fragmentary and even if this were not so, there would be serious modelling problems, because the number of state variables involved would be very large.

Obviously, it is necessary to limit the goals of systems analyses drastically to proceed at all. Rather than analysing all aspects, a distinction is often made between growth and morphogenesis: growth being the main subject of study and morphogenesis being taken more or less for granted. For instance, it is assumed that maize plants develop out of maize seeds, wheat plants out of wheat seeds and spiders out of spider's eggs. In models, such broad assumptions are made operative by introducing preconceived information on the development of the species. For instance, in a model of a wheat plant, a germination, a vegetative and generative stage are distinguished a priori, and it is assumed that in general 9 to 11 leaves develop in the vegetative stage, and that the main growing point develops into the reproductive organ. Likewise, an a priori distinction is made between the successive development stages of insects, as there are eggs, instars, pupae and imagos. What is left to be simulated is the growth within various stages and the rate of their development to subsequent stages. Biologists that are interested in understanding the development of form and function may have another view on the matter and may argue that this approach is too simplified, but appreciation of simplification is more a matter of goal than of principle.

What are the consequences of such an approach for the technique of modelling? Rather than modelling a system fully in terms of measurable state-variables, it is also characterized by historical information which in its most elementary form becomes a record of age only. This is an external record, because age can be known only when the moment of birth is recorded and cannot be determined as such by means of analyses. On the other hand, when age is recorded, relevant properties may be derived from it by correlation.
For instance, in demographic studies the chances of marriage, childbirth, and death may be arrived at in this way. Individuals are lumped at their birth in age-cohorts. Then the ages of the cohorts is kept track of and from them the number of offspring and deaths in a year is calculated. Such a crude technique may do for warm-blooded animals, but not for plants, insects and many other organisms, since their development rate depends largely on environmental conditions. It is then often attempted to conserve past experience in another variable of state: some physiological age. This may be a simple external integral of the temperature: the temperature sum, but also a numerical characterization of the development stage.
As long as such cohorts are characterized by age only, no dispersion occurs. Human individuals that are classified at their birth in the cohort 1970 remain there for their whole lifespan and if nobody is

classified in the cohort 1971, this cohort will remain empty. But, as soon as a physiological age criterion is introduced, some individuals that are born early may age slowly and may be overtaken by individuals that are born later. In other words, individuals that belong to the same age-cohort may become dispersed over a range of physiological ages and it is necessary to develop programming techniques that account for such dispersion phenomena.

This all may seem sophisticated, but such modelling is still very primitive, because it relies in essence on correlations between relevant variables and an external record of past experience and avoids the problem of modelling the main aspects of development of form and function on basic principles.

6.2 Physiological age and development stage

The development stage of warm-blooded animals may be often characterized by a record of the chronological age only. This situation is completely different for many other organisms, such as insects and plants.

Temperature is then often the main determinant, so that the development stage is often accounted for by means of the temperature sum:

```
TS=INTGRL(0.,AMAX1(0.,T-TT))
```

in which T is the current temperature and TT a threshold temperature below which the development processes proceed at a negligible rate. Based on experimental results, it is then assumed that certain development stages are reached at certain values of the temperature sum. For instance, it may be found that the threshold value for maize is 12 degrees centigrade and that tasselling occurs at a temperature sum of 400 degree-days and the plant ripens at a temperature sum of 700 degree-days.

If this approach is taken, it is implicitly assumed that the development rate of the species is proportional to the temperature above the threshold value. However, in general, there is also a non-linear response of development rate to temperature in the higher ranges, as is illustrated in Fig. 18 for two plant species. Here, a constant temperature during growth is given along the horizontal axis and the development rate along the vertical axis, the latter being defined as the inverse of the number of days from emergence to flowering or tasselling.

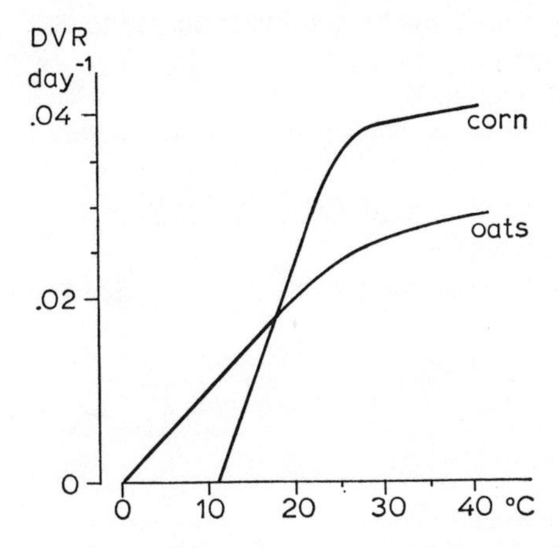

Fig. 18 | The development rate of the plant species maize and oats in relation to temperature at a daylength of 14 hours.

A more sensible approach seems therefore to consider the development stage of the plant as defined by

```
DVS=INTGRL(0.,DVR)
```

in which the development rate in day^{-1} is a function of the current temperature according to

```
DVR=AFGEN(DVRTB,TEMP)
FUNCTION DVRTB=(0.,0.),(12.,0.),...
  (26.,0.035),(28.,0.038),...
  (30.,0.039),(40.,0.041)
```

flowering or tasselling being reached when DVS passes the value of one development unit.

It is assumed that the influence of temperature on the development rate is the same during the whole period of growth, and this assumption is confirmed by the well-known fact that at constant temperature the time between appearance of successive leaves is constant (de Wit et. al., 1970) and that accordingly, a certain calculated development stage fully characterizes the number of leaves and other morphological properties of the plant. The temperature sum or the development stage approach being used, the question remains whether the response in rate of development is immediate or not; that is whether the temperature with its fluctuations throughout the day and from day to day should be used or some average temperature over one day or more. This may make a considerable difference.

Exercise 50

Calculate manually the development stage of maize after 20 days when

a the temperature is 14 °C all the time,

b the temperature is 7 °C for 12 hours of the day and 21 °C for the other 12 hours,

c the temperature is 30 °C all the time,

d the temperature is 40 °C for 6 hours of the day and 26.7 °C for the other 18 hours.

Explain the difference.

At least for plants (de Wit et al., 1970) there are many indications that the response to temperature is instantaneous, so that use of average daily temperatures may lead to considerable errors.

Of course, there are other problems. The development rate may be influenced by daylength or even rate of biomass growth. Like temperature, these factors may also be accounted for on an experimental basis. However, interactions are often so complicated that the development of the plant cannot be accounted for by a simple physiological age. Then more than one characteristic for the development stage may be considered. But problems can then multiply at such a disastrous rate that it is better to take the hard road: modelling of the morphogenesis processes.

What has been said in this section holds in principle for other plant growth stages and other organisms as will be shown later by means of various examples.

6.3 Demographic models

6.3.1 *Age-classes*

Decay of radioactive material occurs with a constant relative rate, apart from random effects that become manifest at low rates. Similar decay processes were assumed to exist for protozoa. However, this is more the exception than the rule with living organisms. In general these organisms develop and age accordingly, and their chances of

dying appear to increase with increasing age.
To simulate such situations it is necessary to have the age distribution of the population at hand. Now it is practically impossible and for most applications unnecessary to memorize the age of each individual. Instead it suffices to memorize the number of individuals in age classes. For instance, in demographic studies it is customary to classify human beings according to their age in years. This is of course an arbitrary choice, depending on purpose. For some applications it would be better to classify according to age in months and for others it would suffice to classify in units of five or ten years.
Such age distributions are memorized conveniently by using the INDEX feature to create a series of age-classes. For instance, human populations may range in age from 0 to about 100 years and if an age distribution has to be stored in age classes of 5 years, it suffices to write:

```
H1=INTGRL(HI1,-PUSH*H1/DELT)
H'2,20'=INTGRL(HI'2,20',PUSH * ...
  (H'1,19'-H'2,20')/DELT)
PUSH=IMPULS(5.,5.)
PARAMETER HI'1,20' = (20 data)
```

Printed output of the variables H1,H2...,H20 may be requested by

```
PRINT H'1,20'
```

Here PUSH has the value zero, except once every 5 years, when its value is set to 1. Only at that moment are the contents of all age classes shifted to the next one. As in other examples (Section 5) this shift is achieved by introducing a rate of change which is equal to the shifted amount divided by the time-interval of integration. The whole age distribution of the population is stored in this way with a resolution of 5 years.
Based on this principle, a simulation program will be written that computes the growth and age distribution of a population with death and birth rates depending on age.
The number of individuals in the first age-class is given by

```
H1=INTGRL(HI1,TBR-PUSH*H1/DELT-RDR1*H1)
```

The total birth rate is the sum of the birth rates from each age-class.

These are given by

```
BR'1,20'=H'1,20'*RBR'1,20'
```

The relative birth rates are given on a parameter card:

```
PARAMETER RBR '1,20' = (20 data)
```

The twenty values are summed with

```
TBR= BR1 + BR2 + .............BR19 + BR20
```

The number of individuals in the other age-classes are given by

```
H'2,20'=INTGRL(HI'2,20',PUSH*...
  (H'1,19'-H'2,20')/DELT-RDR'2,20'*...
  H'2,20')
```

as in the integral for the first age-class.
The total population and the cumulative age-distribution of the population is now found by adding the number of individuals in each age-class with the following recursive operation:

```
TH1=H1
TH'2,20'=TH'1,19' + H'2,20'
```

In demographic studies it is often customary to report relative birth and death rates as a yearly total rather than as an instantaneous rate and then it is best to integrate with time-intervals of one year.

Exercise 51
What is the difference between relative death and birth rates reported as a yearly total rather than as an instantaneous relative rate? Why is it necessary to integrate according to the METHOD RECT?

6.3.2 *Errors of approximation*

The lumping of populations into age classes introduces errors of approximation. These are small and negligible when many classes are used, but may be worth considering if a limited number of classes are distinguished. For instance, in a demographic model of a human population, age-classes of 0–5, 5–10, 10–15 years may be distinguished. Every five years the contents of the classes are shifted one place, so that generally the residence time in each class is five years. This is, however, not so for the first class, because it has a continuous inflow from the birth rate. Only the individuals born just after a shift will stay here five years. As time proceeds the residence time of individuals born later will become progressively shorter. On the average the residence time in the first class will be the half of the 'interval of pushing'. In other words, just after each shift the first age-class contains only individuals close to zero years, and just before the next shift the individuals are 0–5 years. The next age class contains just after the shift, individuals of 0–5 years and just before the next shift individuals of 5–10 years. With a constant birth rate, the average age of the individuals in the age-classes is therefore not 2.5, 7.5, 12.5 years and so on, but 1.25, 5, 10 years and so on. This leads to the conclusion that the age-classes lie between 2.5–7.5, 7.5–12.5 years and so on. The first class covers then the period between -2.5 and $+2.5$ years. Since birth occurs at zero years, the average age in this class is 1.25 years.

There is still a pitfall in initialization. At time zero, each age-class will be initialized with the number of individuals that are between the above given boundaries. Then it takes only 2.5 years before the centre passes to the next class. Therefore the first push should not occur after 5 years, but after 2.5 years, which can be achieved with

```
PUSH=IMPULS(2.5, 5.)
```

Another error is best illustrated by considering the integral for the first age-class, under the assumption that the total birth rate (TBR) is zero for some time. At the time when PUSH = 1, this integral is diminished by its own content and by the number of deaths during that time-interval so that at the next moment the content of the integral is $-\mathrm{H} \times \mathrm{RDR} \times \mathrm{DELT}$ rather than zero. The reason is that too many individuals were shifted. The number that die during this time-interval, should not be removed another time by shifting. Therefore it is necessary to shift not the whole content of the integral but its content minus the number that is lost by death during that time-interval. This is done by writing:

```
H1=INTGRL(HI1,TBR-H1*RDR1-PUSH*...
  (H1/DELT-H1*RDR1))
```

for the first age-class and similar expressions for the others.

Exercise 52
The following tables contain demographic data of the population of the Netherlands on 31 December 1968. The data are grouped in classes with their centres at 1.25, 5, 10,... years (Set 1) and 2.5, 7.5, 12.5,... years (Set 2)

Population size

Class centre in years		Number of men		Number of women	
Set 1	Set 2	Set 1	Set 2	Set 1	Set 2
1.25		305 000		291 000	
	2.5		611 000		582 000
5		612 000		584 000	
	7.5		613 000		587 000
10		597 000		570 000	
	12.5		580 000		553 000
15		575 000		548 000	
	17.5		569 000		543 000
20		576 000		548 000	
	22.5		583 000		554 000
25		517 000		487 000	
	27.5		452 000		420 000
30		429 000		400 000	
	32.5		405 000		380 000
35		399 000		380 000	
	37.5		393 000		381 000
40		382 000		379 000	
	42.5		371 000		378 000
45		367 000		377 000	
	47.5		362 000		376 000
50		338 000		353 000	
	52.5		314 000		330 000
55		306 000		327 000	
	57.5		297 000		323 000
60		280 000		310 000	
80					

	62.5		262 000		298 000
65		223 000		262 000	
	67.5		184 000		226 000
70		184 000		226 000	
	72.5		184 000		226 000
75		150 000		180 000	
	77.5		120 000		150 000
80		90 000		110 000	
	82.5		60 000		70 000
85		40 000		60 000	
	87.5		20 000		25 000
90		13 000		23 000	
and more	92.5 and more		3 000		13 000
Total			6 383 000		6 415 000

Death rates per thousand men and woman per year.

Class centre years		Men		Women	
Set 1	Set 2	Set 1	Set 2	Set 1	Set 2
1.25		15.6		11.4	
	2.5		3.9		2.8
5		1.8		1.2	
	7.5		0.7		0.8
10		0.5		0.3	
	12.5		0.5		0.3
15		0.5		0.3	
	17.5		0.6		0.3
20		0.7		0.4	
	22.5		0.9		0.4
25		1.0		0.4	
	27.5		1.0		0.5
30		1.2		0.6	

	32.5		1.4		0.8
35		1.5		1.0	
	37.5		1.8		1.2
40		2.2		1.5	
	42.5		3.1		2.0
45		4.0		2.5	
	47.5		5.2		3.2
50		6.5		4.0	
	52.5		7.8		4.7
55		9.0		5.5	
	57.5		10.7		6.7
60		11.5		8.0	
	62.4		13.7		10.5
65		16.0		13.0	
	67.5		25.5		16.5
70		35.0		20.0	
	72.5		52.0		35.0
75		70.0		50.0	
	77.5		110.0		85.0
80		150.0		120.0	
	82.5		200.		180.
85		300.		250.	
	87.5		400.		380.
90		600.		500.	
	92.5		900.		760.

The relative number of births per year per age group of the mother.

Class centre years Set 1	Set 2	Set 1	Set 2
15		0	
	17.5		0.022
20		0.091	
	22.5		0.137

25		0.159	
	27.5		0.188
30		0.152	
	32.5		0.113
35		0.084	
	37.5		0.055
40		0.036	
	42.5		0.016
45		0.010	
	47.5		0.002
50		0	

The ratio between the number of boys and number of girls that are born, is 1.048.

Write a simulation program for the growth of the population in the Netherlands, using age-cohorts of 5 years. Which set of data must be used, Set 1 or Set 2?
Why is the time-interval of integration a half year?
Simulate over a period of 50 years and ask for the total men and woman and the relative composition of the population as to sex and age every five years.
Determine also the number of graves after 50 years, if these are maintained for a period of 50, 25 and 10 years.
Death rates during the first year are much higher than during the next years. Is there a simple way of taking this into account?

6.3.3 *The matrix method*

In case DELT in a program with age-classes equals the length of the class, the contents are shifted every time-step one place and are diminished at the same time by the amount died. If the relative death and birth rate do not change with time a matrix method, introduced by Leslie (1945) may be applied to predict the relative composition and the relative growth rate of the population in the stationary state. This is not a simulation method, but will be discussed here because it shows the advantages and disadvantages of matrix algebra versus simulation in demographic studies.

Let the contents of the age-classes be the elements of a vector. If there are ten age-classes, the vector is ten dimensional. The number of individuals in each age-class one time-interval later is now found by multiplying this vector by a matrix as in Fig. 19.

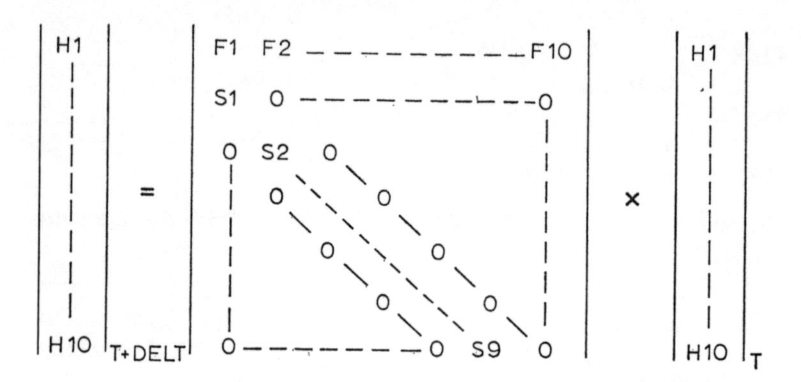

Fig. 19 | The matrix method. H1–H10 are the number of individuals in the age-classes, F1–F10 the relative number of births in and S1–S9 the fraction of each class that survive the time-span DELT.

At the right side, the vector at time T and at the left side the vector at time T+DELT is given. The latter is found by multiplication of the vector at time T by the matrix. In the matrix, FI is the relative number of births per time step in class I and SI is the fraction of class I that passes to I+1; in other words 1 minus the fraction that dies during a time-interval.

It has been proven that repeated multiplication of a vector by a matrix results, in due course, in a vector that has a constant relative composition and whose length increases by a constant factor each time. These are called the dominant eigenvector and the corresponding eigenvalue.
It follows from this that the population will approach a stable age-distribution with a constant relative growth rate, provided that birth and death rates are constant. The standard method to find the dominant eigenvector and its corresponding eigenvalue is the power method

(Faddeev and Faddeeva, 1964) in which the multiplication is repeated until a stable relative composition is reached. This method is therefore very similar to the simulation method and has no computational advantage.
However, there are shorter methods to achieve the eigenvector for matrixes that contain zeros except in the top row and one diagonal. One of these methods is presented in Fig. 20, which is de Jonge's modification of the method of Gauss-Seidel (pers. commun.). This iterative method is very cheap in terms of computing time. It is not explained here because it requires some knowledge of matrix algebra.

```
TITLE MATRIX METHOD APPLIED TO GROWTH OF THE NETHERLANDS POPULATION
PARAMETER C1=1.,R=7.,P=1.
INITIAL
PARAM F'1,17'=3*0.,.055,.343,.47,.282,.137,.04,.005,7*0.
PARAM S'1,16'=.965,.996,.998,.998,.997,.996,.995,.993,.988,.98,.97, ...
      .96,.94,.92,.8,.5
      N=0.
*     ITERATION
NOSORT
    4 CONTINUE
      N=N+1.
      IF(N.GT.20.) GO TO 6
      C'2,17'=C'1,16'*S'1,16'/P
      Q1=F1*C1
      Q'2,17'=Q'1,16'+F'2,17'*C'2,17'
      Q=Q17/C1
      WRITE(6,800)Q,C1,C2,C3,C4,C5,C6,C7,C8,C9,C10,C11,C12,C13,C14, ...
      C15,C16,C17
  800 FORMAT(1H ,F8.5/,9F8.5/,8F8.5///)
      IF(ABS(P-Q).LT.1.E-6) GO TO 6
      P=(R*P+Q)/(R+1.)
      GO TO 4
    6 CONTINUE
DYNAMIC
TIMER FINTIM=1.,DELT=1.
END
STOP
ENDJOB
```

Fig. 20 | An iterative determination of the eigenvector and its corresponding eigenvalue of a matrix as in Fig. 19, written as an INITIAL section in CSMP.

The method gives directly the eventual stable age-distribution and the corresponding relative growth rate, which is the eigenvalue minus one. The method does not give the total population after n years. To achieve this important value, the power method or straight-forward simulation must be applied.

6.4 Germination models

6.4.1 *Boxcar train without dispersion*

Like the development of plants, the germination of seeds or the hatching of eggs may take some time, which depends on environmental conditions, especially temperature. If a certain amount of seeds is placed suddenly in a position where the germination process may start, its germination stage at any moment may be defined by

```
GS=INTGRL(0.,VDV)
```

in which the velocity of development in day^{-1} is defined as a function of temperature by

```
VDV=AFGEN(VDVTB,TEMP)
FUNCTION VDVTB=(10.,0.065),...
  (15,0.143),(20.,0.143)
```

The data hold for seeds of the winter annual *Veronica arvensis*, that have been stored for 15 weeks (Janssen, 1973).

Exercise 53
Write a simulation program for the germination stage, in which the temperature varies sinusoidally with the time of day with an amplitude of 5 degrees, and an average of 15 degrees (see also Fig. 2). The computation may be terminated as soon as the germination stage passes the value 1. What does this mean? How is this achieved?

The above procedure may be used to follow the development of one batch of seeds. However, it is easy to vizualize a situation with seeds in different stages of germination and then their age-distribution has to be taken into account. For this purpose, classes have to be distinguished and because development is very much a function of temperature, these must be development classes rather than age classes. Hence, the contents must not be shifted at preset time-intervals, but at the moments that the development stage is increased by the inverse of the number of classes (N).

Exercise 54
Why 1/N?

This is achieved by defining a 'PUSH' according to

```
PUSH=INSW(GS-1/N,0.,1.)
GS=INTGRL(0.,VDV-PUSH*1/N/DELT)
```

Here PUSH is set to one by the INSWitch, at the moment GS is larger than 1/N. This moves the contents of the classes and decreases at the same time the integral GS by the amount 1/N, resetting this integral at the correct value close to 0. GS is increased again at the proper rate by the velocity of development.

Exercise 55
Write now a simulation program for the germination of *Veronica arvense* seeds at 20°C with 10 development classes. Execute the program introducing an initial amount of 1000 seeds at time zero. At which moment do these seeds germinate?

6.4.2 *Boxcar train with constant relative dispersion*

Usually germination does not take the same number of days for different seeds, because neither the seeds nor their micro-environments are exactly the same. The overall effect is illustrated in Table 5, where the percentage germination of a batch of *Veronica arvense* seeds is given.

Table 5 Germination percentages of *Veronica arvense* seeds at 10°C, stored for 15 weeks.

Day	10	13	14	15	16	20	22	27
percent germ.	1	12	28	46	56	87	91	100

Exercise 56
Make graphs of the percentage germination and the rate of germination against time. Calculate the average time of germination and its standard deviation from the data of Table 5.

The time curve for the rate of germination has the bell-shaped form of the Gaussian distribution function.
It will be shown that such distribution functions are obtained also by simulation, if the contents of the classes are not pushed at certain moments but moved continuously from one development class to the next with a rate that is proportional to the rate of development. In the most simple situation, only one development class is considered—ungerminated seeds—and germination is described as an exponential decay process of ungerminated seeds according to

```
H=INTGRL(HI,-RTG)
```

in which the rate of germination is given by

```
RTG=H*RDV
```

and RDV is the relative rate of development, or germination.
The total amount of seed that are germinated equals then

```
TG=INTGRL(0.,RTG)
```

H,RTG and TG are presented in Fig. 21. For obvious reasons (see also Chapter 2) H and RTG decrease exponentially with time and TG approaches HI accordingly. The average germination period is the integral of the rate of germination at any moment multiplied by

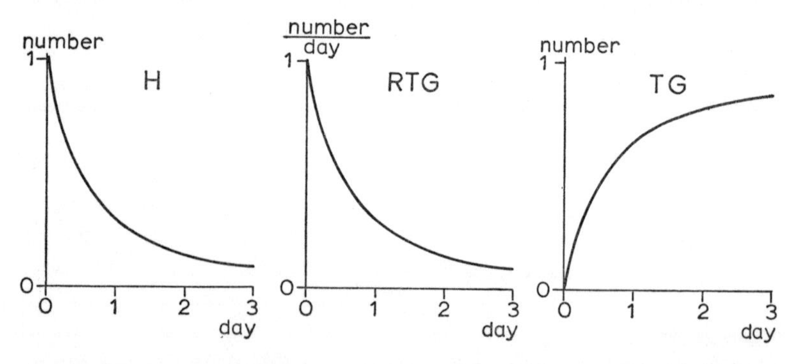

Fig. 21 | Amount of ungerminated seeds (H), germination rate (RTG) and amount of germinated seeds (TG), when germination is described as an exponential decay of ungerminated seeds with a relative germination rate of 1 day^{-1}.

the time that has elapsed since the start of the process, standardized at a unit amount of seed (HI = 1). This is the standardized surface under the curve of H versus time in Fig. 21 and in CSMP notation defined by

```
AGP=INTGRL(0.,TIME*RTG/HI)
```

Only the last value, when H is decreased to practically zero, is right.

Exercise 57

Calculate manually the average germination period when out of a batch of 100 seeds:

100 germinate on day 5,
100 germinate on day 10,
50 germinate on day 5, and 50 on day 10
75 germinate on day 5, and 25 on day 10

Finish the simulation program to calculate the average germination period. Use the method RKS for integration and terminate simulation as soon as the content of H is 1/100 of its original content. Execute the program for a relative rate of development of 0.01, 0.05, 0.1 and 0.5 day^{-1} and take FINTIM equal to 500 days. Multiply the average germination period by the relative germination rate. What is the dimension of this product AGP × RDV and what is its numerical value?

Prove now the equality of the inverse of the relative germination rate and the average germination period by making use of the analytical expression:

$$H = HI \times e^{-RDV \times T}$$

and of the equality:

$$\frac{1}{HI}\int_0^{HI} T \times dH = -\frac{1}{HI}\int_0^{\infty} T \times \frac{dH}{dT} \times dT$$

If the above exercise is done properly it will be clear that the product of the average germination period and the relative rate of development (APG × RDV) is always 1. Hence the relative rate of development as defined in the above program may be replaced by the inverse of the average germination period.

The results that are obtained by considering only one development class of ungerminated seeds describe much more a decay process of seeds than a germination process. This is different when more development classes are considered, as is again most conveniently done by means of the INDEX feature. For instance, a germination process that is described by means of 10 development classes may be programmed as follows:

```
H1=INTGRL (HI, H1/REST)
H'2,10'=INTGRL(0.,(H'1,9'-H'2,10')/REST)
TG=INTGRL(0.,H10/REST)
```

Obviously, when the average germination period is AGP and the number of classes 10, then the residence time (REST) in each class is

```
REST=AGP/10
```

Exercise 58
What is the time constant (TAU, as defined in Section 3.5) of this system?

The average germination period may be again a function of the environmental conditions and the time-interval of integration should be a tiny fraction of REST.
The rate of germination and the cumulative amount of germinated seeds—the breakthrough curve—are given in Fig. 22 by the curves marked 10, it being assumed that the average germination period is 20 days. The curves marked 5 and 20 holds when 5 and 20 development classes are distinguished. The form of the curve suggests that the simulation procedure leads to a Gaussian distribution function of germination, at least when a sufficient number of classes are used. With low class numbers, the results suggest a Poisson distribution. A detailed mathematical analysis (Goudriaan, 1973) shows that this is indeed so and that the relation between residence time (REST) or the average germination period ($AGP = REST \times N$), the number of classes (N) and the standard deviation of germination (S) is given by

$$N = S^2/REST^2 = AGP^2/S^2. \tag{6.1}$$

provided that the time-interval of integration (DELT) is small enough. This relation gives the number of development classes, that are

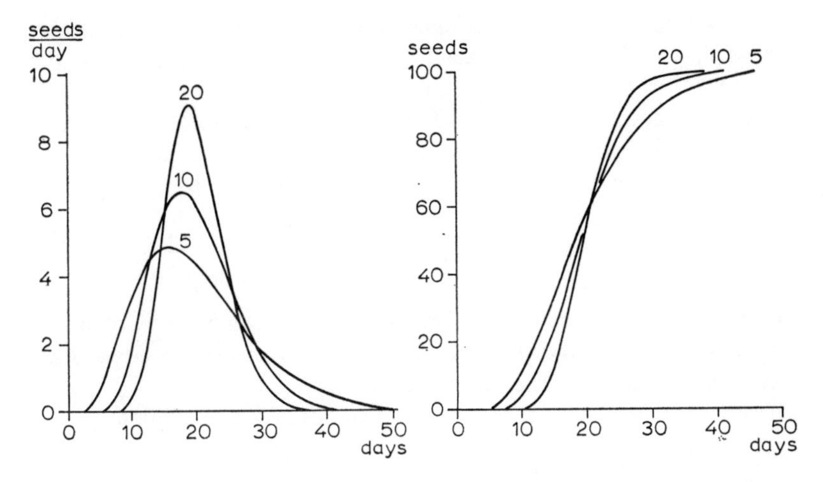

Fig. 22 | The rate of germination and the breakthrough curve, when 5, 10 and 20 development classes are considered and the average germination period is 20 days.

necessary to achieve a certain relative dispersion (S/AGP), independent of the average germination period.
The method is rather flexible. It is not necessary to start with a given batch of seeds and death rates depending on conditions may be introduced at any development stage. Moreover the transfer of contents is continuous, so that the method RKS with self-adapting time-interval of integration may be used.

6.4.3 *Boxcar train with controlled dispersion*

Two methods to simulate germination have been discussed. The first method does not introduce any dispersion and the second method gives a constant relative dispersion, once the number of development classes is fixed. There are, however, a few remaining problems. In the first place, the number of classes is already 100 when a relative dispersion of 10 percent is to be simulated and in the second place, it is impossible to change the relative dispersion according to conditions, because the numbers of classes cannot be varied during simulation.

Both problems may be overcome by following an intermediate course, in which the fraction F of the contents of each class is shifted at the fraction F of the residence time in a boxcar.
This is programmed as follows for 10 development classes.

```
H1=INTGRL(H1,-H1*PUSH*F/DELT)
H'2,10'=INTGRL(0.,(H'1,9'-H'2,10')*...
  PUSH*F/DELT)
PUSH=INSW(GS-1.,0.,1.)
GS=INTGRL(0.,1./(F*REST)-PUSH/DELT)
```

Inspection of the statements shows that no dispersion is obtained when F equals 1 and a constant relative dispersion, as defined by Eqn (6.1) when F is set equal DELT/REST.
It can be shown that for intermediate values of F the relation

$$N = \frac{AGP^2}{S^2} \times (1-F) \qquad (6.2)$$

holds.
With F equal to DELT/REST and DELT sufficiently small, this equation transforms, of course, into Eqn (6.1).
Figure 23 gives an example of the result. The continuous curve is

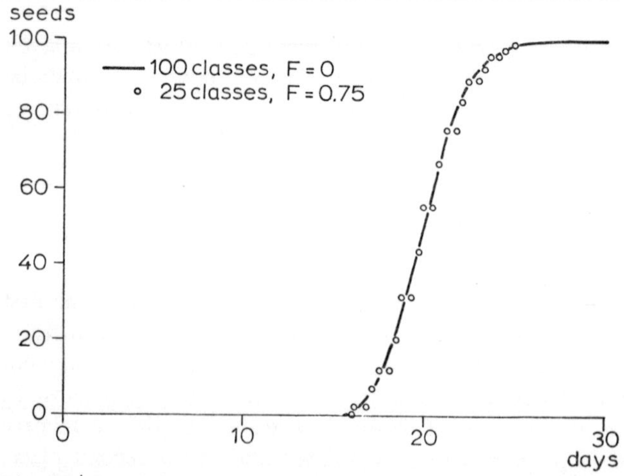

Fig. 23 | Breakthrough curves for 100 development classes with F equal to zero and for 25 development classes with F equal to 0.75. The average germination period is 20 days.

obtained by means of 100 development classes and F equal to DELT/REST. The dots are the result of using 25 classes and F equal to 0.75. In both cases the relative dispersion is 0.1, but in the second case there is not a smooth curve. The given dots have OUTDEL as time-interval. The discontinuity and the use of METHOD RECT is the penalty that has to be paid for reducing the number of classes and retaining a small dispersion. As has been said, the advantage of the procedure is that also F, and with this the dispersion, can now be varied independent of the average germination period and the number of classes that has been chosen.

Exercise 59

What is the value of F when the relative dispersion is 0.25 and N equals 25? What should be done in this situation?

This method with controlled dispersion has been used by Janssen (1973) to simulate the germination of *Veronica arvensis* and *Myosotis ramossima* seeds. However useful, it should be realized that by applying this type of simulation, results of experiments are 'mimicked' rather than simulated. The term mimicked is used here to emphasize that the main aim is the summarizing of the experimental results in a program that simulates germination, but that no serious attempts are made at this stage to base the equations and parameters that are used on more detailed physiological knowledge of the processes involved.

Exercise 60

Complete the following table for F = 0.5:

TIME	H_1	H_2	H_3	H_4
0	1	0	0	0
0.5 × REST				
1.0 × REST				
1.5 × REST				

in which REST is the residence time in each class. What is the name of the resulting probability distribution function?

This function was discussed in Section 5.6. Readers with some knowledge of probability calculations should read this section again and answer the following questions.

Express B and f of Section 5.6 in TIME, F and REST.

What kind of probability distribution is obtained when F equals DELT/REST and DELT approaches zero?

6.4.4 *Errors of approximation*

A discussion on the demographic models showed that the age-class that was intended to cover for instance the years 10–15, appeared to cover the years 7.5–12.5; the average age of the class was half its range lower. For continuous flow, this error of lumping does not occur, but the simulation process results in a constant, relative dispersion.

The simulation method with controlled dispersion ranges between two situations: when $F = 1$, the error of lumping is fully present and when $F = DELT/REST$ (and DELT small) the error is absent. It can be derived that for any value of F the shift in development of each class equals $F \times REST/2$. Therefore, in front of the first class a 'pre-class' is constructed with an average residence time of $F \times REST/2$, so that the centres of the following classes are independent of the value of F.

This is achieved by

```
H0=INTGRL(0.,RTIN-RTOUT)
RTOUT=H0*2./(F*REST)
H1=INTGRL(HI1,RTOUT-PUSH*F*H1/DELT)
H2=...........
PUSH=INSW(GS-1.,0.,1.)
GS=INTGRL(0.5,1./(F*REST)-PUSH/DELT)
```

The continuous inflow `(RTIN)` enters `H0` rather than `H1`. This correction must not be applied to the initial amounts, so that the initial value of `H0` is always zero and only `H1,H2,...` are initialized. The initial value of `GS` is set at 0.5, so that the first `PUSH` occurs after `0.5 * REST`, in this way the initial average age or development within each class is correctly accounted for.

Exercise 61

Apply this method to simulate the growth of the Netherlands population. Which set of data out of Exercise 52 should be used now?

6.5 The flow of heat in soils

There are considerable similarities between the simulation of ageing and dispersion in populations and of physical diffusion and dispersion processes in time and space. The similarities will be illustrated here by developing a simulation program for the flow of heat and temperature variations in the soil with the temperature at the surface as a forcing function.
For this purpose a uniform soil column from an infinite slab is considered which is placed on an insulating layer. To calculate the temperature as a function of time and depth, this column is divided into 25 equal compartments. Heat flow into and out of each compartment is calculated at any instant of time from the temperature difference between compartments and the conductivity between compartments. These heat flows are integrated to follow the heat content of each compartment and thus the temperature.
Simulation is done most conveniently by creating integrals of the heat content via:

```
HC'1,25'=INTGRL(HCI,NFL'1,25')
```

If the soil is uniform, the compartments are of the same size (TCOM) and the initial temperature (TI) does not vary with depth, then the initial heat content is given by

```
HCI=TCOM*VHCAP*TI
```

in which VHCAP is the volumetric heat capacity of the soil. The net flow into each layer is the difference between the flows over the boundaries:

```
NFL '1,25' = FLW'1,25'-FLW'2,26'
```

Exercise 62
Which direction of flow is assumed to be positive?

The flow is proportional to the temperature differences between the layers and the conductivity of the soil (COND) and inversely proportional tot he distance between the centres of the layers (here also TCOM):

```
FLW'2,25'=(TMP'1,24'-TMP'2,25')*COND/TCOM
```

The flow out of the 25th layer is zero, because the column is placed on an insulating layer. It would also be zero if the column was taken so long that temperature changes in the last compartment were negligible. Hence:

```
FLW26=0.
```

The flow into the first layer is

```
FLW1=(TMPS-TMP1)*COND/(0.5*TCOM)
```

in which the temperature at the surface has to be defined as a forcing function, for instance:

```
TMPS=TAV + TAMPL*SIN(6.28*TIME)
```

if a cyclic daily fluctuation is assumed.

Exercise 63
Why is the thickness of the compartment multiplied by 0.5?
What is the unit of time?
What are TAV and TAMPL?

The temperature of the compartments is obtained by:

```
TMP'1,25'=HC'1,25'/(TCOM*VHCAP)
```

The integration is best done with

```
METHOD RKS
```

and a stationary state of the cyclic variations is obtained in about 4 days, so that

```
TIMER FINTIM=4, PRDEL=0.1
```

suffices.
The output of all 25 temperatures and of other relevant parameters are requested with

```
PRINT TMPS, FLW1, TMP'1,25'
```

As an example, the parameters are defined with:

```
PARAMETER TCOM=2.,COND=360.,...
  VHCAP=1.05, TI=20.
```

with time in days, distance in cm, heat in joule and temperature in °C.

Exercise 64
What are the units of all variables and parameters used in the simulation program?

With a uniform soil and with a sinusoidal forcing function, the variation of temperature may be calculated also by means of an analytical solution. This has been done for comparison, the result being presented in Fig. 24. It appears that the analytical and simulated solution agree

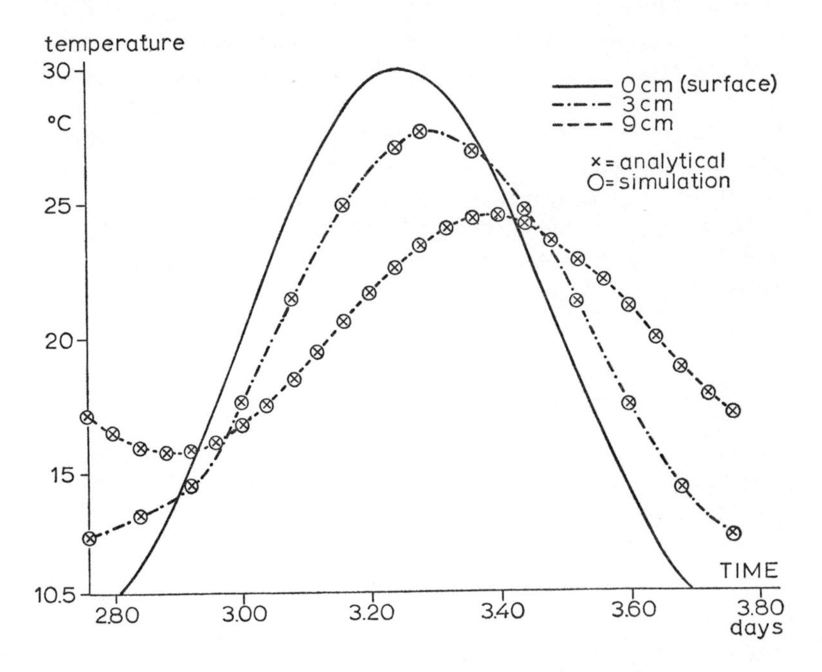

Fig. 24 | Analytical and simulated solution for the temperature course in a uniform soil, with a sinusoidal temperature variation at the surface.

within 0.01 percent and this shows that it is not necessary to use very thin compartments for accurate results.

Exercise 65
Rewrite the program, so that conductivities and heat capacities that vary with depth can be introduced. Is it necessary to use compartments of the same size?

In another monograph of this series (de Wit and van Keulen, 1972), simulation programs of this type have been developed to study also the transport and diffusion of water, salts and ions in soils, the only difference being that instead of the INDEX feature, the more cumbersome DO-loop feature from FORTRAN is used. In other studies (Goudriaan & Waggoner, 1972), similar techniques are also used to simulate micro-meteorological phenomena, but it goes beyond the scope of this monograph to discuss the principles of these.
At last it is remarked that simulation programs with compartmentalization of space may be used to study dispersion of animals, seeds and spores. However, it should be remarked that these may become large, when dispersion in two and certainly in three dimensions is considered, so that other techniques must be developed for these purposes.

7 Growth and development of Helminthosporium maydis

7.1 Introduction

Helminthosporium maydis is a fungus of corn *(Zea maize)*. Especially the leaves may be covered with lesions, which develop microscopic sporophores. These produce spores, that are dispersed by wind and rain and so reach new healthy leaf tissue. There they germinate and penetrate the plant tissue; new lesions appear after incubation. Under suitable conditions, the life cycle is completed within a week.
The fungus is responsible for Southern corn leaf blight, a disease that, especially in 1970, ravaged the corn fields of the USA. The yield was 15 percent less than that estimated before the disease struck, and losses of half or more were common in the Gulf region. The disease suddenly appeared because the T (Texas) type of cytoplasmatic male sterility was applied on a large scale in the hybrid system. This type appeared vulnerable for *H. maydis*, which had existed for a long time in a non-virulent form.
To anticipate the growth of the disease in the field, Waggoner et al. (1972) analysed this new disease and made a simulation program for its growth and development. A comparison of important aspects of simulated results with field observations (Shaner, Newman, Stirm & Lower, 1972) showed the merit of this approach.
The simulation program for the growth and development of epidemics of *H. maydis* (EPIMAY) is written in FORTRAN and keeps track of the development of the lesions formed on each day after infection. This makes the program difficult to read and grasp. A further analysis of Waggoner & de Wit showed that a simulation program that is much more lucid and easier to handle could be developed by using the state variable approach as developed in this monograph.
The meteorological factors that effect the fungus are temperature, light, wetness, wind and rain. The influence of these factors on growth and development of the disease in various life cycles was analysed for the Illinois isolate of race T. of *H. maydis* growing on

the corn cultivar PA 602 A, F1 hybrid in the laboratory and the greenhouse. Undoubtedly the condition of the host affects the growth responses, but the study was restricted to well fertilized and good growing specimens of the host. These observations and general knowledge about growth and morphogenesis of the fungus form the basis for the construction of the simulation program. This program will be presented in the form of relational diagrams, together with sufficient quantitative information to leave the writing of the actual program to the reader of this monograph.

7.2 The weather

Even, if a corn crop is uniform, the micro-meteorological conditions for developing fungus are not the same, but vary with height. The radiation during the day is higher, the wind more turbulent and the leaves are dry longer near the top of the crop than near the soil surface. Programs to simulate the micro-meteorological conditions in the crop are still being developed (Goudriaan & Waggoner, 1972) but these are likely to be of use only after the simulation program for the pathogen is refined. At present the microclimate in the crop is not simulated, but instead the macro-weather factors are employed as forcing functions as some 'average' for the whole crop.

Exercise 66
Why is this a dangerous approach?

The parameters are temperature, wind speed, light, rain and the presence of water on the leaves. These can be introduced in the form of function data throughout a season, but for the present it suffices to define a particular daily course of the weather which is repeated every day. The following weather data are assumed for some simulations in this chapter.

```
FUNCTION TEMPT = (0.,14.),(12.,35),...
  (24.,14.)
FUNCTION WINDT = (0.,1.),(6.,1.),...
  (14.,4.),(19.,2.),(24.,1.)
FUNCTION WETT = (0.,1),(7.99,1),...
  (8.,0.),(19.99,0.),(20.,1.),(24.,1.)
```

```
FUNCTION LITET = (0.,-1.),(5.99,-1),...
  (6.,1),(20.,1.),(20.01,-1.),(24.,-1.)
FUNCTION RAINT = (0,0),(24,0)
```

The units for temperature, wind and rain are °C, m s^{-1} and mm h^{-1}, respectively. Especially the temperature course is simplified, to facilitate later analyses of the results. For light and wetness only two conditions are distinguished:
Light (`LITE = 1`) and dark (`LITE = -1`) and wet (`WET = 1`) and dry leaves (`WET = 0`).
To read the graphs, time in hours during the day has to be known. As `TIME` is expressed in days, the hour of the day may be calculated with

```
HOUR = 24*(TIME-AINT(TIME))
```

in which the function `AINT(TIME)` conserves the integer part of time, and assumes, for instance, the value 6 when time is between 6 and 7 days.
Although it does not belong to the weather section, the growth of the crop must be considered. Simulation of a disease is especially important when crop growth is not seriously affected, because that is the time to control the disease. Thus we can assume that crop growth is independent of the growth of the disease, so that it can be introduced in the program as another forcing function. It suffices to use for this purpose the course of the leaf area index, that is the ratio between the surface of the leaves and the surface of the soil, which varies from 0 at emergence to about 5 at flowering. In the present simulation it is simply assumed that

```
FUNCTION LAIT = (0.,3),(140.,3)
LAI = AFGEN(LAIT,TIME)
```

Exercise 67
Write the section `WEATHER` of the simulation program, complete with `AFGEN` functions and `FUNCTION` tables and the temperature (`TEMP`), the wind speed (`WIND`), the wetness of the leaves (`WET`), the dryness of the leaves (`DRY`), the light condition (`LITE`) and `LAI` as outputs.

Mistakes in input data may result in a situation where it rains and

WET is nevertheless zero. Inconsistencies may be avoided by reading from the tables an auxiliary variable WETX and then computing

```
WET = FCNSW(WETX + RAIN, 0.,0.,1)
```

which means that WET = 1 for WETX + RAIN greater than 0, and otherwise 0.

7.3 Appearance and growth of lesions

If spores of *H. maydis* are present on healthy leaves and conditions for germination are suitable, some spores will form germ tubes which penetrate through the stomata and so infect the leaves. This penetration rate will be calculated at the end of the program and used here as an input.

Fig. 25a shows the resulting growth of the lesion area at 30 °C as a function of the number of days after incubation. Lesions appear after about 2 days showing that the first stages of development occur inside the leaves. Thereafter the lesions grow to their final size with

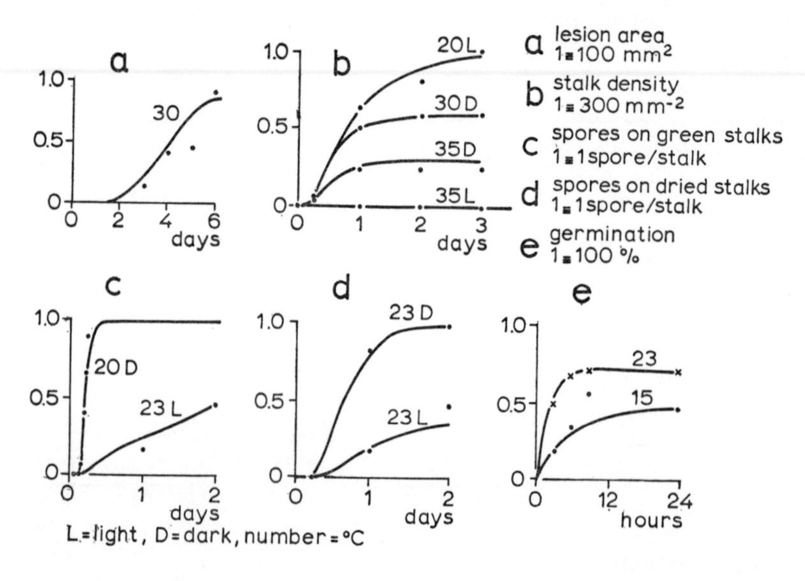

Fig. 25 | Some experimental data and the curves mimicked by the relevant parts of the simulation programs.

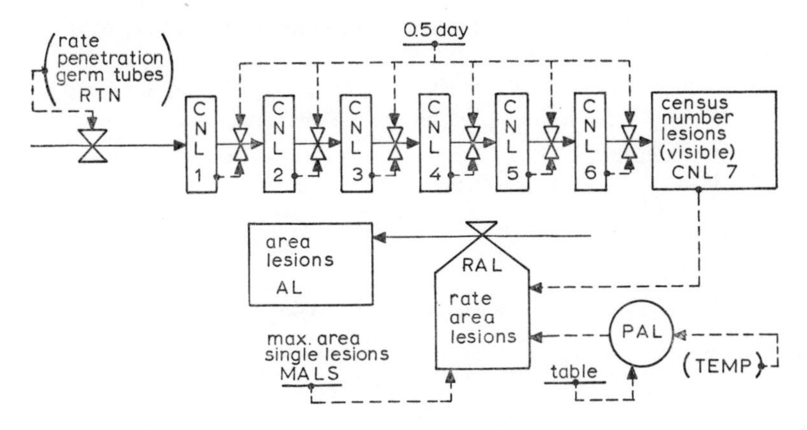

Fig. 26 | Relational diagram for the growth of lesions.

a speed that is dependent on temperature only.

The relational diagram of development and growth of the lesions is given in Fig. 26. The development period inside the leaves is accounted for by 7 development classes with dispersion, and a residence time of 0.5 day in the first 6 classes. This residence time is assumed to be independent of temperature. The content of the last integral gives the census of the visible lesions. As long as it is assumed that defoliation and decay of leaves are negligible, this number does not decrease. Since, a soil surface of one hectare is used as reference, the census of lesions is expressed in ha^{-1}. All lesions grow to a final size (MAL) of about 100 mm^2 or 10^{-8} ha and the growth rate of the individual lesions can be conveniently described by assuming that this rate is proportional to the difference of the maximum area of a lesion minus its actual area ALS. The proportionality factor (PAL) is, according to experiments, a function of temperature only, and is sufficiently defined by

```
FUNCTION PALT = (0,0.),(10,.14),...
  (18,.33),(23,.80),(30,.80),(35,.14),...
  (40,.0) day⁻¹
```

The formula for the growth rate of the area of a *single* lesion is then:

```
RALS = PAL*(MALS-ALS)
ALS = INTGRL(0.,RALS)
```

The initial value of this integral is zero, because the lesions entering class CNL7 have an area that is practically zero.
To obtain the growth rate of the *total* area, RALS must be summed over all the visible lesions present, a number equal to CNL7:

$$\text{RAL} = \sum_{n=1}^{\text{CNL7}} \text{PAL} \times (\text{MALS}-\text{ALS})$$

or

$$\text{RAL} = \text{PAL} \times (\text{CNL7} \times \text{MALS} - \sum_{n=1}^{\text{CNL7}} \text{ALS})$$

or

```
RAL=PAL×(CNL7×MALS-AL)
```

if AL is the total area of the lesions, given by

```
AL=INTGRL(0.,RAL)
```

Exercise 68
Why is the expression for RAL so similar to the one for RALS? Write the section GROWTH OF LESIONS, with the number of visible lesions CNL7, the rate of growth of the total area, and the total area as outputs. What is the total residence time of the lesions in the invisible stages. Explain why some lesions are already visible at 1.5 day. Calculate the standard deviation of lesion appearance. Why has the simulated curve for AL in Fig. 25a a sigmoid form?

The points in Fig. 25 are observations and the curve is the mimicked result. A similar analogy between observation and simulation is obtained at other temperatures.

7.4 Sporophore or stalk formation

The technical term for the microscopic stalk that holds the spore in the air above the leaf is sporophore, but here the more popular term 'stalk' will be used. The growth of the stalk occurs only when the leaves are wet and otherwise depends on temperature and light. The maximum number of stalks on a hectare of lesions is 300×10^{10}, but the experimental data in Fig. 25b for a few temperature and light conditions during formation show that this maximum is not reached under all conditions. Moreover there appears to be some delay in the formation of stalks.

To mimick these results it is assumed that there is a potential number of stalks per surface unit of lesions—a number of opportunities for stalk formation—which materialize through some classes and that during the actual stalk formation a part of this potential number develop into stalks and the rest become extinct, depending on conditions. These assumptions are presented in the relational diagram of Fig. 27. The growth of the number of opportunities is the product of the maximum number per area (MOA $= 300 \times 10^{10}$ per ha) and the growth of the area of the lesions. This potential number enters

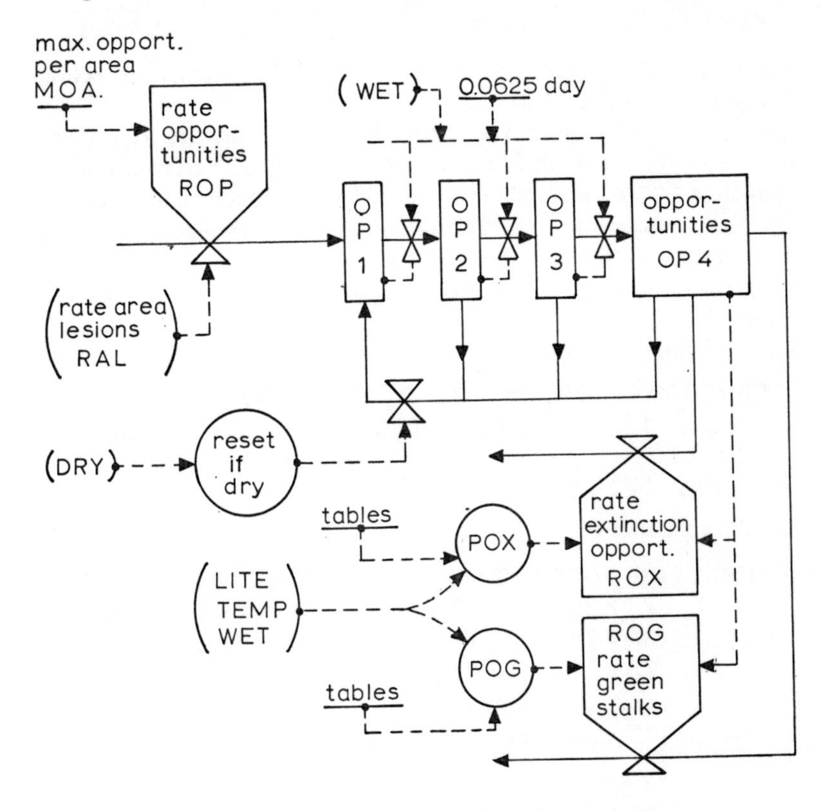

Fig. 27 | Relational diagram for the formation of green stalks.

into a series of 4 classes with a residence time of 0.0625 day in three classes. The realization of the opportunities is arrested by drought. There are three possibilities: opportunities are destroyed, set back to their initial stage or their advance is arrested. Not much is known about this, so that at present the middle course is taken: it is assumed that the opportunities are returned to the first class in case of drought. This can be done by introducing the following rate out of the integral equation of the Xth class

```
EMPT=DRY*OPX/DELT
```

To avoid manipulation of very small numbers in some computers ('underflows') it may be advisable to program this rate as

```
EMPT=INSW(OPT-1.E-50, 0., DRY*OPX/DELT)
```

Hence the classes are not emptied when their contents are below the very small value of 10^{-50}.
The opportunities end in the last or fourth class and are from there removed either by stalk formation or extinction. The relative rate of extinction (POX) and stalk formation (POG) depend on temperature and light, whereas the process only occurs when the leaves are wet. An analysis of the experimental data showed that the process is sufficiently mimicked when the following functions of temperature are used:
During light:

```
FUNCTION POGL = (0,0),(14,.04),...
  (18,.12),(23,1.4),(30,1.2),(35,0) day-1
FUNCTION POXL = (0,0),(14,.04),...
  (18,.12),(23,1.4),(30,0) day-1
```

and during darkness:

```
FUNCTION POGD=(0,0),(14,.10),(18,.27),...
  (23,.27),(30,1.33),(35,.67),(40,0) day-1
FUNCTION POXD=(0,0),(14,.02),(18,.03),...
  (23,.18),(30,.88),(35,1.54),(40,0) day-1
```

The proper functions can be selected again by an inswitch which is operated by the variable LITE. For instance:

```
POG = INSW(LITE,AFGEN(POGD,TEMP),...
  AFGEN(POGL, TEMP))
```

The points in Fig. 25b are again observational data for a few conditions and the corresponding curves are obtained by mimicking the process of stalk-formation and opportunity extinction. It should be realized that the process of opportunity formation is described by the equations and the functions, but not explained. It is therefore not feasible to explain the form of the functions on a physiological basis. The stalks that are formed are virgin or green stalks. Because these maintain another rate of spore formation than stalks that sporulated once or were subjected to drought, they have to be accounted for separately in an integral that maintains the census of green stalks.

Exercise 69

Write now the section FORMATION OF GREEN STALKS, with the rate of green stalks formation (ROG) as output. What is the dimension of ROG? This rate as a fraction of the potential rate (sum of actual formation and extinction) depends on light and temperature. What are these fractions when the stalks are formed in the light at 21 and 32 °C and in the dark at 18 and 27 °C both on wet leaves?

7.5 Sporulation of green stalks

The name 'green' stalks has been used explicity because there are also 'dried' stalks. Dried stalks are stalks that have sporulated at least once or have been subjected at least once to drought. The distinction is made because the influence of temperature and light on sporulation is different for both categories: green stalks sporulate more rapidly than dried stalks.

Fig. 25c shows how this growth of spores on green stalks may depend on temperature and light. Here the scale of 0 to 1 represents the number of green stalks with a spore. A stalk cannot carry more than one spore at the same time. Only 50 percent of the stalks produced spores after two days in the light and at 23 °C, but there is sufficient information to assume that in due course all stalks will sporulate under these conditions.

The relational diagram for sporulation of green stalks is given in Fig. 28. Three classes with a residence time of 0.0625 day in the first two are again introduced to mimick the observed delay between the formation of green stalks and the first appearance of spores. The first

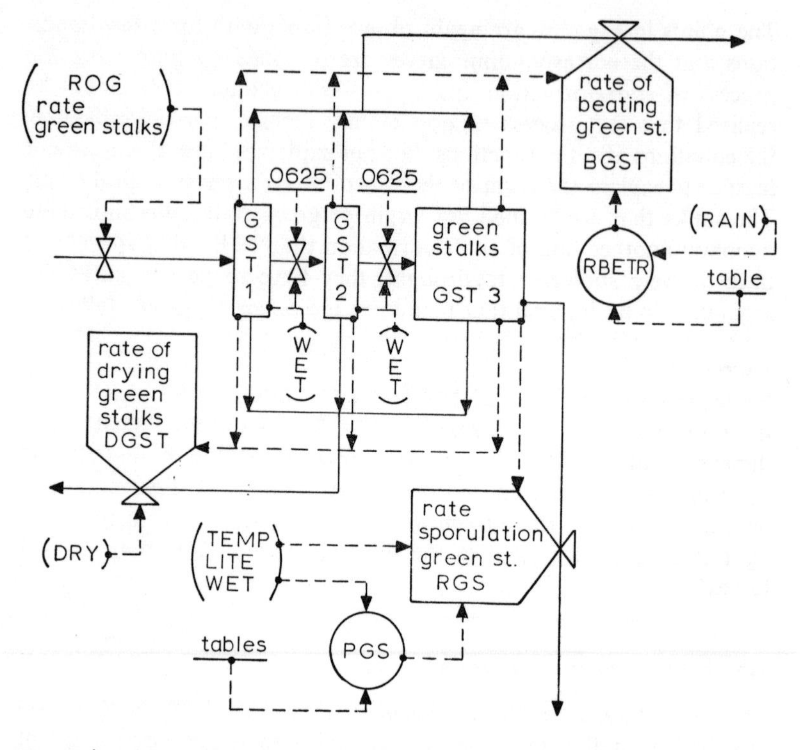

Fig. 28 | Relational diagram for the formation of spores on green stalks.

class is loaded according to the rate of green stalk formation; thus this class contains stalks ready for sporulation.

There are two circumstances that arrest spore formation. One when the leaves become dry; then the growing spores are aborted and the green stalks are reclassified as dried stalks. The other when the green stalks are completely destroyed, usually by rain beating against the fragile stalks. The relative destruction rate is assumed to be a function of the rainfall rate, in mm hour^{-1} according to

```
RBETR=AFGEN(BEATT,RAIN)
FUNCTION BEATT= (0,0),(0.25,.08),...
  (0.75,.32),(6.25,2),(18.8,5.6),...
  (25.,6.7) day-1
```

This function summarizes some factual information, but is largely based on a qualified opinion of the process.

Exercise 70

What rate of rainfall is needed to destroy 63 percent of the stalks in 5 hours?

When the stalks have been passing through the classes and have not been dried up or beaten by rain, they form spores at a rate which is dependent on light and temperature, provided, of course, that the leaves stay wet. It appeared that the experimental data are sufficiently accurately mimicked by introducing the temperature dependency.

FUNCTION PGSL = (0,0),(14,.15),...
(18,1.44),(23,.32),(30,0),(40,0) day^{-1}

in the light and

FUNCTION PGSD=(0,0),(14,.06),(18,14),...
(23,14),(30,.44),(35,0),(40,0) day^{-1}

in the dark for the proportionality factor of spore formation. The points in Fig. 25c are again observations and the curves mimick results of sporulation of green stalks.

Exercise 71

Write now the section FORMATION OF SPORES ON GREEN STALKS with the rate of spore formation on green stalks (RGS) as output. What is the dimension of RGS?

Further spore formation is arrested, as long as the spore remains on the stalk. Once removed, the stalk is no longer green, but classified as a dry stalk, which may also form spores but at a different rate.

7.6 Sporulation of dried stalks

As has been said, dry stalks are distinct from green stalks because their rate of spore formation is slower. Dried stalks are generated in various ways. When spores are removed from either a green or a dried stalk, the stalk is ready to produce a new spore at a rate characteristic for dried stalks. This is also the case when during spore formation the growing spore is aborted by drought. Fig. 25d shows

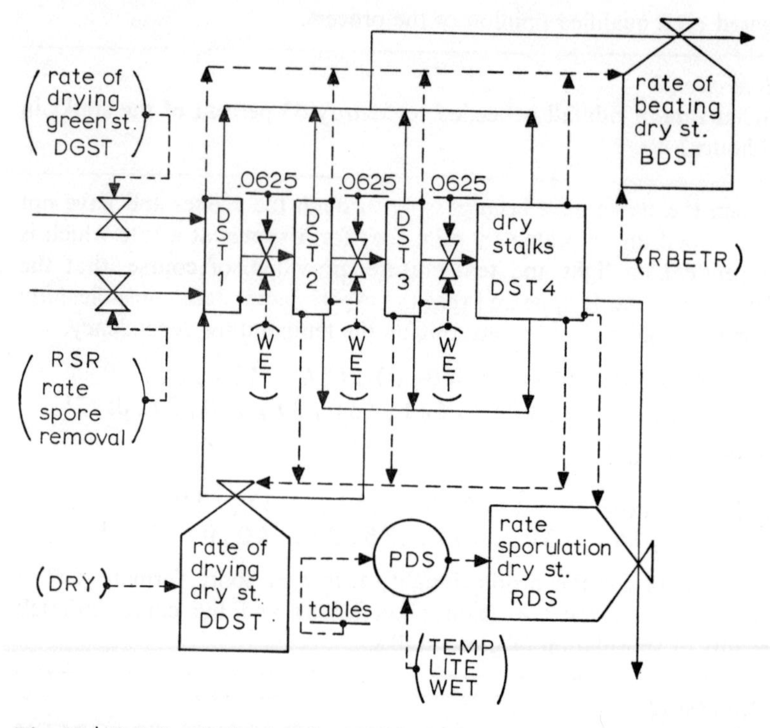

Fig. 29 | Relational diagram for the formation of spores on dried stalks.

some experimental results which are mimicked according to the relational diagram in Fig. 29. The mean residence time in each class is again 0.0625 day, the slowness of the process as compared to green stalks being accounted for by an extra class and another proportionality factor for spore formation according to the temperature dependency

```
FUNCTION PDSL = (0,0),(14,.17),...
  (18,1.75),(23,.25),(30,0),(40,0)
```
day^{-1}

in the light and

```
FUNCTION PDSD = (0,0),(14,.07),...
  (18,2.95),(23,2.2),(30,.53),(35,0),...
  (40,0)
```
day^{-1}

in the dark.

When dried, the growing spores are aborted and the dried stalks are again reset into the first class. When the green stalks become dry they also enter this class. At last, the stalks that are denuded of spores, either by wind or rain and not destroyed in the process are again ready to form new spores. Dried stalks are also beaten and destroyed by rain at the same rate as for green stalks.

Exercise 72
Write now the section FORMATION OF SPORES ON DRIED STALKS with the rate of spore formation on dried stalks (RDS) as output. Inputs are the rate of transfer of green stalks and dried stalks to the first class during drought and the rate of spore removal from stalks (RSR).

7.7 Dispersal of spores

When it is dry, the spores are removed by the turbulent action of the air. Some of the spores are carried away to other fields, and others settle on the soil, on lesions or on healthy foliage segments. The stalks are also denuded by rain. Especially at the onset of heavy showers, part of the spores are dispersed through the air, but with gentler rain the spores are washed from the stalks and end up again on the soil, on lesions or on healthy foliage. The processes that are involved are very little understood particularly because the quantitative aspects are complicated: fields may be of limited size and infections are not uniformly distributed.
At such stages, the model builder has to make a difficult decision: either to abandon the whole problem or to advance for better or for worse. The latter course is usually chosen for various good reasons. First, sensitivity analyses may show that the dynamics of the system are hardly determined by the processes that are difficult to handle both conceptionally and practically. Thus, it would be a waste of time to pay much attention to these processes. Unfortunately, this appears not to be the case here: spore dispersal is one of the important processes that governs fungal epidemics. Secondly, life goes on and operational decisions have to be made whether the system is completely understood or not: even models with unsatisfactory parts may be better than no model. Of course, this has to be proven. Thirdly, it is

possible to view a model not so much as a representation of the real system but as a representation of our knowledge of the system and our opinion about it. Then the weak sections should not be ignored but exposed and this will be done here.
The most simple supposition is that spores are removed from stalks at a rate proportional to the number of spores present. The relative rate of removal is assumed to be zero when the leaves are wet and it does not rain. However, when it rains the spore removal rate is

```
SPRR = RWASH * STSP
```

and when it is dry

```
SPRD = RBLOW * STSP
```

in which STSP is the integral 'stalks with spore'.
The relative rates of spore removal (RWASH and RBLOW) are assumed to be independent of the number of stalks with spores, although it is not unlikely that these relative rates decrease because at first the most exposed spores are removed. The main problem is to obtain a reasonable estimate of these relative rates.
Waggoner et al. estimated that with a sprinkling rate of about 6 mm/hour for 3 hours, 86 percent of the spores were removed from exposed leaves. This means that the value of RWASH $= -8.\ \ln(0.14) = 15.7$ day^{-1}.

Exercise 73
Prove that this is the case.

It does not seem unreasonable to assume that RWASH is proportional to the rainfall intensity and that below an LAI of 2 the leaves do not protect each other, so that RWASH is independent of LAI. Above this value mutual protection may exist, but since water may drop from one leaf on to the other, no large mistakes may be made when this protective effect is neglected.
The relative rate of spore removal under dry conditions depends primarily on the wind velocity. It was assumed by Waggoner et al., that at a wind speed of 2 metres per second, and a leaf area index of 3, about 5 percent of the spores are removed in 3 hours, so that SPRD can be estimated under this circumstance. Since the force of the wind is proportional to the square of its velocity, it

could be assumed that the relative rate of spore removal is proportional to the second power of the wind-speed. Then spore removal is zero at zero wind-speed. However, turbulence is also generated by the temperature difference within and outside the crop. This effect may be approximated by assuming that the wind-speed is never less than 1 m/s. The relative rate of spore removal is also influenced by the leaf area index, because the wind-speed decreases more or less exponentially with increasing depth of the crop. This effect is so uncertain, that it is not considered further.

Exercise 74
Calculate RBLOW for WIND equal to 2 m s^{-1}.

This completes the estimates of the rate of spore removal. The next step is the estimation of the fraction of removed spores that may become effective by settling on healthy foliage.
With strong winds and a small field most spores may be blown away and become ineffective. However, they may be compensated for by spores blown in from neighbouring fields. Another question is how many spores in the air are caught by leaves and how many end up on the soil surface where they can do no harm. Again Waggoner et al. assumed that with a leaf area index of 3 and a wind-speed of 2 m sec^{-1}, 3 percent of the spores are caught by the leaves. This percentage is likely to depend more or less linearly on the leaf area and is programmed as such. The percentage is also likely to decrease with increasing wind-speed, especially on small fields. This effect is too complicated to consider here.
The greater the intensity of the rain, the more spores are washed to the ground. According to Waggoner et al. only 0.3 percent of the spores are caught by the leaves at a rainfall intensity of 2.5 mm/hour. The maximum fraction is caught at a negligible rainfall rate, but does not exceed 20 percent. These are very rough estimates indeed.

Exercise 75
Write the section SPORE DISPERSAL with as output: SPRR, SPRD and their sum RSR and the rate of spore arrival at the foliage (RASP). How would it be possible to take into account the influence of host exhaustion when the area of the lesions is not negligible?

7.8 Germination of spores and penetration of germ tubes

The spores on healthy tissue are now considered. These germinate eventually on wet foliage. The fraction of spores that complete germination and the rate of germination depend on temperature according to the observations in Fig. 29e. The process of germination is complicated because germinating spores may be washed down from the leaves or killed upon desiccation.

The relational diagram that describes germination and penetration of germ tubes into the leaves is presented in Fig. 30. Two integrals are distinguished: the census of spores on the (healthy) foliage (CSF) and the census of germ tubes on (healthy) foliage (CGT). The census of spores increases because of arrival of new spores (RASP) and decreases because spores are washed, killed or germinate. The relative rate of spore removal by rain from the leaves is set equal to the relative rate of spore removal by rain from the stalks (RWASH). The killing of spores upon desiccation is more difficult to handle.

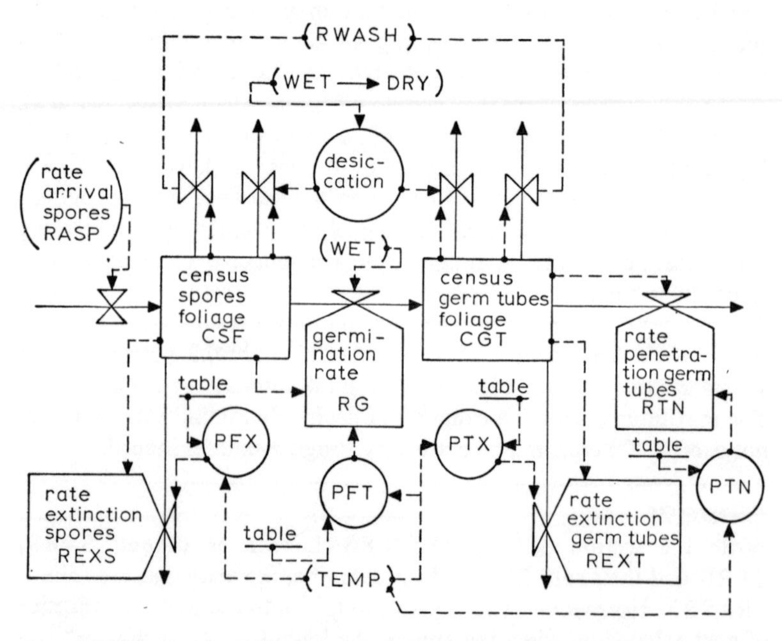

Fig. 30 | The relational diagram for germination and penetration of spores.

Spores can only desiccate when they have been wet. Since all spores are killed, the killing rate is

```
RKSP = KILL*CSF/DELT
```

Obviously, the killing is governed by the variable KILL which may be 0 or 1. If the leaves during the previous time-interval were wet and are dry during the current time-interval, KILL is set at 1. Such a condition may be programmed by using a 'PROCEDURE' that contains a series of statements that have to be executed in the order in which they are presented. The whole sequence of statements is then sorted at a place where the inputs are available and the outputs not yet used.
The procedure that is used here is called 'DESS' from dessication and has as input the variable WET and as output the variable KILL:

```
PROCEDURE KILL = DESS(WET)
```

The statements within the procedure are:

```
KILL=0.
IF((WETP-WET).GT.0.)KILL=1.
WETP=WET
```

The first statement sets KILL equal to 0 and the second statement reads: if the difference between WETP and WET is greater than 0, then reset KILL to the value 1. The next statement sets the previous value of wet (WETP) equal to the current value and this reset value is used in the 'IF' statement during the next updating. The end of the series of statements that have to be sorted as one block is now defined with the line

```
ENDPROCEDURE
```

The spores germinate or become extinct at relative rates that depend on temperature, according to the function tables

```
FUNCTION PFTT=(0,0),(10,.4),(15,1.8),...
  (20,4.6),(23,7.0),(35,3.7),(40,0) day-1
```

for completion and

```
FUNCTION PFXT=(0,0),(10,0),(15,1.8),...
  (20,4.2),(23,2.6),(35,3.7),(40,0) day-1
```

for extinction.

The simulated germination is again presented by the curves in Fig. 25e. Note that at 15 °C some observations deviate considerably from the simulated line. This is because the function tables PFTT and PFXT are assumed to be smooth and were adapted also to observational data at other temperatures.
The spores with germ tubes are also killed upon desiccation according to the rate

```
RKGT = KILL*CGT/DELT
```

and also washed away by rain at the same relative rate RWASH as spores are washed from the stalk. Depending on temperature, only a fraction of the germ tubes ever penetrate the leaves; this observation is again mimicked by introducing relative rates of penetration and extinction according to

```
FUNCTION PTNT = (0,0),(18,.48),...
  (23,.65),(30,.25),(35,0),(40,0)
```
day^{-1}

for completion and

```
FUNCTION PTXT = (0,0),(18,1.3),...
  (23,2.6),(30,2.2),(35,0),(40,0)
```
day^{-1}

for extinction.
These functions are found by comparing the number of lesions with the number of germ tubes formed upon incubation of spores.

Exercise 76
Write the section GERMINATION AND PENETRATION with the rate of penetration (RTN) as output.

The cycle is completed by calculating the rate of penetration of the germ tubes, RTN being the rate needed to start the growth of the number of lesions.

7.9 Timing, initialization and output organization

Since there are discontinuous processes involved it is necessary to execute the simulation according to the METHOD RECT. The time-interval of integration has to be chosen small compared with the relative rate of changes. An analysis of the data and parameters shows

that these are the fastest in the classes for the growth of the stalks, which are governed by a residence time of 0.0625 day. When DELT equals this value, the contents of the classes are pushed without any dispersion. Here this is completely acceptable.
For practical reasons of organizing input and output it is, however, convenient to set DELT at 0.04 day. Then the program is updated 25 times during one day and computing costs are acceptable.
The initialization of every integral in the program could be achieved by observing at one moment the number and area of lesions, the number of green and dry stalks, the number of spores and so on in a particular field. This is of course not worth the trouble at this stage of knowledge. Usually initialization is achieved by assuming a certain number of spores or a certain number of lesions, the contents of the other integrals being set at zero. Because it is often the purpose to study the dynamics of the disease without complications due to exhaustion of the host, it is good practice to start with a small number of lesions, which may be taken as 100 per hectare.
However, in other situations it may be necessary to program a certain invasion rate of spores from the outside during some period.

Exercise 77

Program an invasion rate of 10^6 spores per hectare per hour during the first week, but only when it is light and the leaves are dry.

The output of every variable may be requested of course, but it is good practice to limit the number to the most essential ones. These are in general the contents of the main integrals: the number and area of the lesions, the number of green and dry stalks and of stalks with spores and the number of spores and germ tubes on healthy leaves. To study the behaviour at certain moments it may be convenient to have all outputs available. This may be achieved by introducing the statement

```
OUT1 = DEBUG (N,T)
```

in which T is the moment at which this output procedure starts to operate and N the number (without decimal point) of successive updates for which output is requested. As many debugs as needed may be introduced.

Exercise 78
Ask for a debug of 10 rounds at time zero and of debugs of 2 rounds at time 5, 5.5, 10 and 10.5.

7.10 Results and sensitivity analyses

A simulated epidemic, as characterized by visible lesion number (CNL7), is presented in Fig. 31 on a logarithmic scale, starting with 100 lesions per hectare (CNL1), the growth being simulated for the defined stationary weather pattern. During the first periods of growth, the effect of initialization can still be distinguished. At a later stage

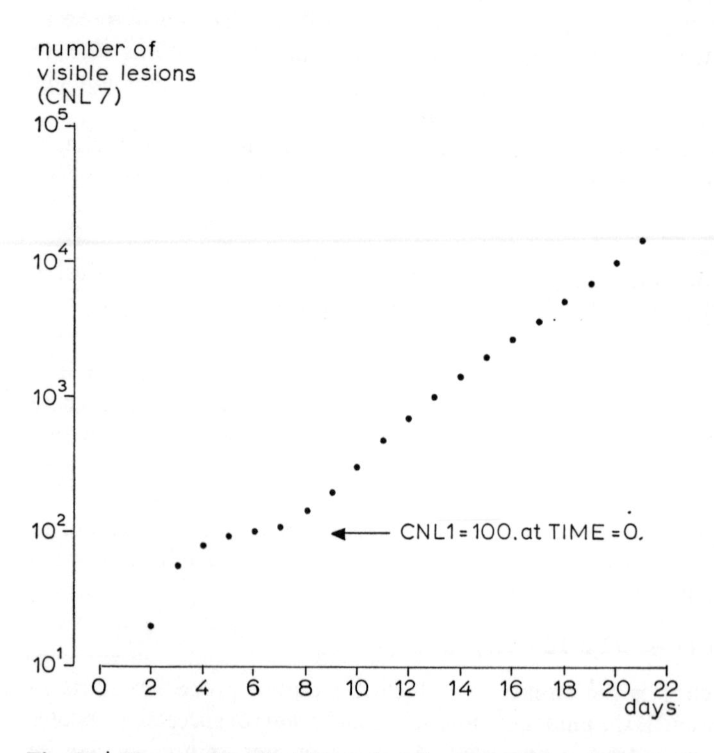

Fig. 31 | The number of visible lesions in dependence of time, when initialized with 100 spores that completed penetration (CNL1 = 100).

it is possible to characterize growth by a relative growth rate of number of fungi lesions, which is in the present example 0.34 day^{-1}. Other important characteristics are the rate of spore production (RSP) and dispersion by wind (SPRD). The simulated results of these for days 8 and 9 are given in Fig. 32, together with relevant weather data. The rate of spore removal by wind may be verified in a relative sense by comparing with the density of spores above the crop.

Verification of simulated data on epidemics is difficult for two reasons. In the first place, a good meteorological network that provides not only the course of the standard meteorological parameters throughout the day, but also detailed information of the wetness of the leaves must be available. In the second place, field observations must be organized. Sometimes a rating of severity in a wide range of localities may do, but preferably the relative growth rate of the disease over

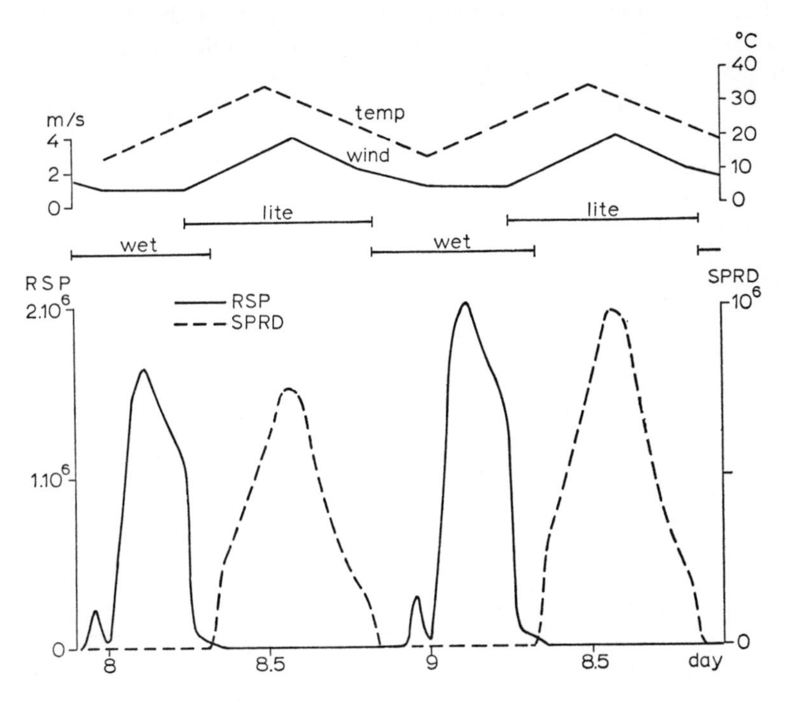

Fig. 32 | The simulated rate of spore production (RSP) and rate of spore removal by wind (SPRD) during two days.

a certain time span should be observed. Some comparisons of simulated results, obtained with the original 'EPIMAY' of Waggoner et al. and actual results throughout the United States are given in Fig. 33. The left graph is a comparison of the simulated multiplication rate of lesions with a net increase of blight ratings in various places in the Mid-Eastern United States in 1971 and the right graph compares simulated and actual multiplication rates in Western Indiana in 1971. Only the latter gives a comparison in absolute terms, but it should be taken into account that some 'fudging' of parameters has been done to achieve correspondence of level. Whether such fudging is acceptable or not is not much a matter of principle, but of purpose. If it is the purpose to develop a forecasting technique as soon as possible one may incorporate experience of previous years into the program. It should be realized, however, that in that case it is very difficult to judge which of the numerous parameters should be left alone and which should be adapted. If much adaption is necessary, it is doubtful whether much is gained at all by simulation compared with the application of one of the standard multiple correlation techniques.

If it is the purpose to understand the dynamics and the quantitative

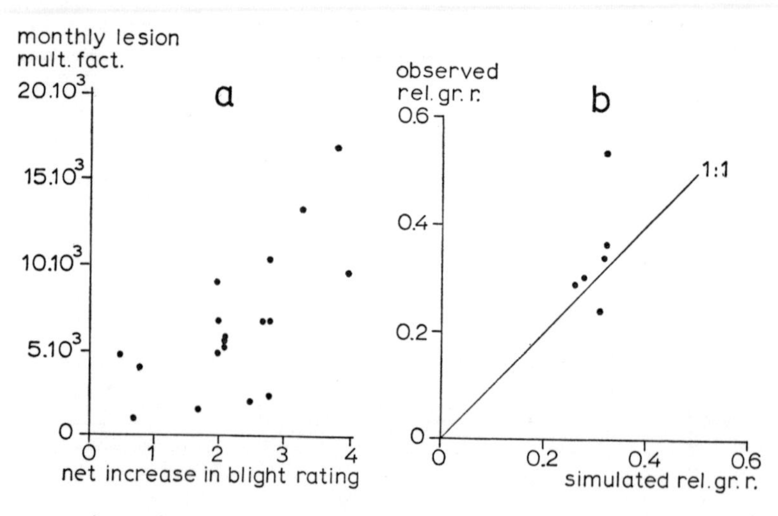

Fig. 33 | A comparison of simulated and actual results in the Mid-Eastern United States (a) and in Western Indiana (b) in 1971.

aspects of the disease, fudging of parameters to achieve better agreement should be avoided. Instead, a sensitivity analysis under the prevailing conditions should be executed, to evaluate which parameters are mainly controlling the disease. The result of this analysis, should then be a guide to further experiments and study.

Such a sensitivity analysis consists of varying inputs and parameters over a certain range and a comparison of their relative influence on the end result. If the influence of a certain parameter or input is relatively small, further analyses may be left for some time, but if the influence is large, more work should be invested in a further analysis of the section of the program where this parameter plays a role.

The weather parameters that are most likely to affect the severity of the epidemic are the duration of the wetness of the leaves, the presence of showers and the temperature. The simulated influence of duration of wetness on the relative growth rate of the lesions for the standard weather conditions, but in the absence of rain, is given in Fig. 34. The propagation of the disease is practically zero when the duration of wetness is less than 4 hours, because the fungus needs wetness periods of finite length to complete its development in various stages. The relative multiplication rate increases to a maximum at 18 hours

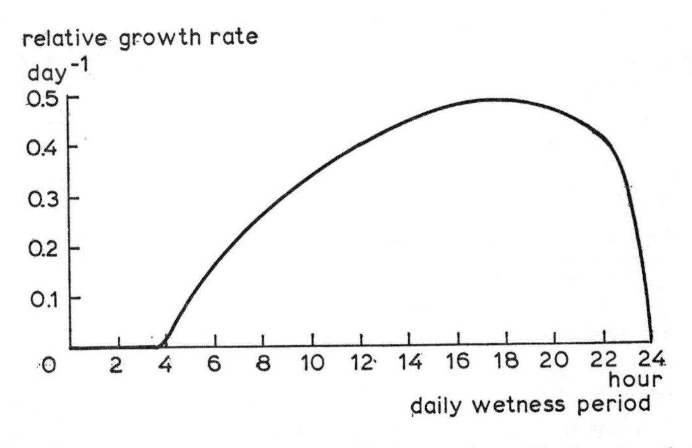

Fig. 34 | The relative growth rate of the number of lesions in relation to the duration of wetness, without rain.

of wetness, but then it decreases again to zero at 24 hours of wetness, because spores are assumed not to disperse by wind when the leaves are wet, and because rain is supposed to be absent.

The picture changes completely when at 24 hours of wetness rain is assumed to occur at a rate of 6 mm per hour for 3 hours per day; then the relative growth rate equals 0.19 day^{-1}. In this case the rain causes the dispersal of the spores. The influence of decreasing the intensity of the shower and increasing its duration is considerable. A rainfall of 1 mm per hour for 18 hours per day causes a relative growth rate of 0.84 day^{-1}. At a lower rate the relative beating rate RBETR decreases so that fewer stalks are destroyed, but the spore dispersal by rain increases, so that many more spores are caught by leaves. As long as the total amount of rain is the same, the change of RWASH has little influence because it is proportional to RAIN. All this means that knowledge of the daily total rainfall is not sufficient; the rainfall distribution must be known as well.

The influence of temperature is analysed under the assumption that the other weather conditions are standard. Two situations are distinguished: in one series, the temperature amplitude is fixed at 5°C and the average temperature is varied from 15°C to 35°C and in the other the average temperature is fixed at 25°C and the amplitude is varied from 0° to 15°C.

Exercise 79

Program this situation by assuming a sinusoidal temperature course throughout the day with a maximum at 14.00.

The results are given in Table 6 and show that one temperature value, such as an average temperature, does not give detailed enough information.

The influence of daylength under otherwise standard conditions is found to be small.

A sensitivity analysis of the parameters and function tables that are included in the program may be made also. For instance the influence of the residence time in the various classes may be evaluated and this the more so because these are assumed to be independent of temperature. Another aspect that may be of importance is the assumption regarding the development of green stalks. Does it make much difference whether developing stalks are destroyed by drought during

Table 6 The influence of temperature average and amplitude (°C) on the relative growth rate (day^{-1})

AVTMP AMPL	15	20	25	30	35
0	—	—	0.452	—	—
5	0.079	0.321	0.459	0.407	0.192
10	—	—	0.389	—	—
15	—	—	0.254	—	—

development, whether their growth is only arrested or whether they are reset in their first class upon drought? It may also be questioned whether it is worthwhile at all to make a distinction between green and dry stalks. Another important and largely unknown set of parameters concerns dispersal and recapture.

Exercise 80

Make a sensitivity analysis of the influence of the parameters that are mentioned and some others that may be considered important. Design experiments to elucidate a better formulation.

8 Solutions of the exercises

8.1 Introduction

1 The differential equations for the falling apple are:

$$\frac{\mathrm{d}v}{\mathrm{d}t} = g, \qquad \frac{\mathrm{d}s}{\mathrm{d}t} = v$$

The rate of change of the amount of electric charge on a capacitor is equal to its charging current and the potential across the capacitor is equal to its amount of charge divided by its capacitance, so that

$$\frac{\mathrm{d}e}{\mathrm{d}t} = i/c$$

By substituting $i_1 = g.c_1$ and $i_2 = e_1.c_2$
for the charging currents of a first and second capacitor, the differential equations for the falling apple are obtained.

2 Our results are as follows:

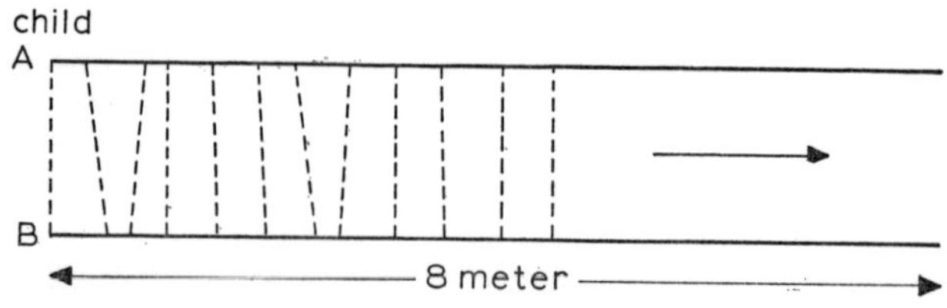

Fig. 35

In a first model of this system it may be sufficient to assume that there are two variables of state, which increase by the rate of walking of each child. There are two independent rate determining processes: one in the mind of each child which wants to stay side by side with the other. However, this model would not explain why the step sizes in the beginning of the process are more variable than at the end.

To understand this it has to be realized that each child has a memory which is able to conserve the sizes of the step of the other. Each child thus determines the step size to be taken not only on the difference in position but also on a 'normal step size for the other child', figured from data conserved in memory. These memorized data characterize also a state of the system. Hence a more sophisticated model requires more than two state variables. Digital computers are much more suitable to memorize such historical data than analogue computers and this is one of the other main reasons why they are preferred to simulate complicated state determined systems.

8.2 Exponential growth

3 The dimension of A is grams and of GR is grams $hour^{-1}$. Eqn (2.1) is then dimensionally consistent when RGR is expressed in grams per hour per gram or in $hour^{-1}$. There should be a constant amount of food and of harmful waste products. This is usually achieved by an abundant food supply beyond saturation, and an entire removal of waste products. The temperature should also be constant.

4 The result is

TIME	0	2	4	6	8	10
A	1.000	1.221	1.492	1.822	2.226	2.718

The relation between the logarithm of the amount and time is linear, since taking the logarithm is per definition the inverse of taking the exponent:
if

$$A = e^{RGR \times T}$$

then

$$\ln(A) = RGR \times T$$

in which ln stands for the logarithm with base e.
It is recalled that

$${}^{10}\log(e) = 0.43429$$

or that

$${}^{e}\log(10) = \ln(10) = 2.3026,$$

so that

$$\ln(X) = 2.3026 \times {}^{10}\log(X)$$

5 The results with DELT = 2 hours are:

TIME	A	RGR × A	RGR × A × DELT
0	1.000	0.1	0.2
2	1.200	0.12	0.24
4	1.440	0.144	0.288
6	1.7280	0.1728	0.3456
8	2.0736	0.20736	0.41472
10	2.4883		

6 Some results with DELT equal to 1 and 0.5 hours are

TIME	0	2	4	6	8	10
A(DELT = 1)	1.000	1.210	1.464	1.772	2.144	2.594
A(DELT = 0.5)	1.000	1.216	1.477	1.796	2.183	2.653

A_0 (= IA) being the initial amount, A equals $A_0 + A_0 \times \mathrm{DELT} \times \mathrm{RGR}$ after one time-interval and after two time-intervals $A_2 = A_1 + A_1 \times \mathrm{DELT} \times \mathrm{RGR}$
In general the relation

$$A_n = A_{n-1}(1 + \mathrm{DELT} \times \mathrm{RGR})$$

holds.
Since A_{n-1} can be written as the product of A_{n-2} and $(1 + \mathrm{DELT} \times \mathrm{RGR})$ and RGR is constant, the expression can be transformed into

$$A_n = A_0(1 + \mathrm{DELT} \times \mathrm{RGR})^n$$

This is the value of A_n at time $n \times \mathrm{DELT}$, so that

$$A_n = A_0(1 + \mathrm{DELT} \times \mathrm{RGR})^{\mathrm{TIME/DELT}}$$

or

$$A_n = A_0((1 + 1/X)^X)^{\mathrm{RGR} \times \mathrm{TIME}}$$

with $X = 1/(\mathrm{DELT} \times \mathrm{RGR})$
When TIME stays constant and DELT approaches zero, X approaches infinity and the expression for A approaches to

$$A = A_0 \times e^{RGR \times TIME}$$

in which the number e is standing for

$$e = \lim_{X \to \infty} (1+1/X)^X = 2.7182 \ldots$$

This is the so-called analytical solution for the differential equation of exponential growth, which is just a standardized way to write the procedure for a numerical solution.

7 RGR is the only variable containing the dimension of time. If RGR is expressed in $hour^{-1}$, TIME assumes the dimension hour. If programs contain more variables with the dimension time, care must be taken to use the same unit of time.
Of course it is always necessary to express variables in the same dimensional units: a practical problem that should not be underestimated.

8 See Fig. 1.

9 The rectilinear method of integration is used and in the solution for Exercise 6 it is shown that the analytical solution is the limit of the numerical solution with DELT approaching to zero. It is always necessary to use a not too small interval DELT in simulation because otherwise round-off errors would accumulate and the computation would never be terminated.

10 Please do not take the trouble to find a mathematical expression for such kind of relationships.

11 Compared with the scatter of the observational data, the deviations between the smoothed curve and the straight segments is small, so that it is unnecessary to use smaller temperature-intervals in the tabulated function.

12 sine (0°) = 0.
sine (15°) = 0.259
sine (30°) = 0.5
sine (60°) = 0.865
sine (90°) = 1.

The maximum temperature is reached just 6 hours from the beginning of the day.

13 The rate of change of T is 1, so that $T_{T+DELT} = T_T + DELT$ and because the initial value of T is zero, T = TIME.

14 The answers do not differ very much for different choices of DELT, so that 0.5 hour seems to be a reasonable choice. The relative growth rate is 0.1 at a temperature of 12.94 °C and the temperature is maintained on this level by introducing

```
PARAMETER AVTMP=12.94, AMPTMP=0.
```

15

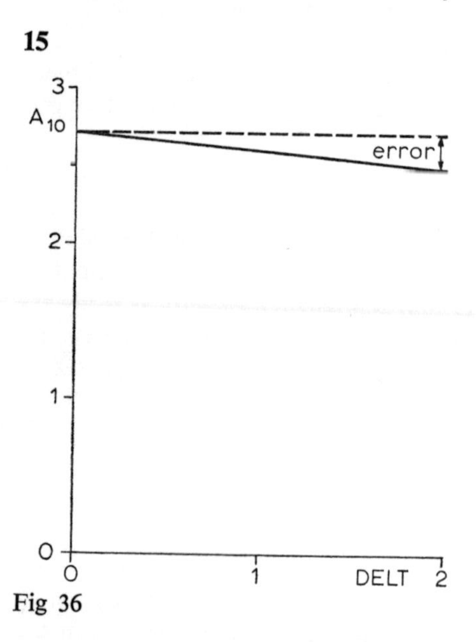

Fig 36

16 It is read from the graph that the maximum size of the time-interval is about 1.2 hours if a relative deviation of 5 percent from the analytical solution is acceptable. This is about 1/10 of the inverse of the relative growth rate. The relative deviation increases with the length of the simulation period. Starting with the correct value at 10 hours, the relative deviation is again 5 percent at 20 hours. These deviations are additive, so that the total relative deviation after

20 hours is 10 percent. With a two times larger RGR, the time scale is compressed by a factor 2, so that the time-interval must be compressed also by a factor 2, to obtain a relative deviation of 5 percent at 5 hours. To obtain this deviation at 10 hours, it is necessary to reduce DELT with another factor of 2. Hence the time interval has to be taken four times smaller when RGR is two times larger.

17 The allowable time-interval is obviously larger than OUTDEL, but OUTDEL is taken because output is requested at these intervals. When using METHOD RKSFX, DELT can be set equal to OUTDEL. The tabulated results for RGR 10 × larger are:

HOUR	RGR	ΔCOUNT
6– 7	2.14	4
7– 8	2.11	4
8– 9	2.04	4
9–10	1.94	4
10–11	1.81	4
11–12	1.67	4
12–13	1.49	2
13–14	1.28	2
14–15	1.09	2
15–16	0.93	2
16–17	0.82	2
17–18	0.76	2

8.3 The growth of yeast

18 It was said that two rates do not depend on each other, but not that one rate cannot depend on the other. Here, the rate of growth and the rate of alcohol production are consequences of the same process: the biosynthesis of yeast material out of sugar. Therefore, there is a fixed ratio between rate of growth and rate of alcohol production. The rate of sugar consumption is stochiometrically related to the above two rates: laws of conservation of matter, energy etc. can be formulated in such a way that some rate of appearance always equals some rate of disappearance.

19 The rate of sugar consumption is equal to a sugar consumption factor times the rate of yeast growth for each species. The amount of sugar is an integral which is emptied by both rates. The amount or concentration of the sugar in the medium should feed back on the growth rate of the yeasts. The quantitative aspects of this feed back are not presented in the diagram.

20 RED=ALC/MALC, since in this case $0 \leqslant ALC \leqslant MALC$. Otherwise, RED should be given by

```
RED=LIMIT(0.,1.,ALC/MALC)
```

21 The best estimate of RGR1 is obtained by presenting the amount of yeast during early growth on a logarithmic scale against time and drawing a straight line through the data. The value appears to be about 0.2 $hour^{-1}$. The value of ALPF1 is the alcohol concentration at the end, divided by the amount of yeast or $1.5/13 = 0.115$ percentage of alcohol per unit of yeast. The alcohol production factor depends on the size of the vessel. In a larger vessel, the same amount of alcohol would cause a smaller percentage. It would be more elegant not to mix up the influence of physiological aspects (alcohol production rates) with experimental aspects (vessel size), but Gause did not give the latter. The alcohol percentage corresponding with the initial amount of yeast is $ALPF1 \times IY1$, but Gause did not add this alcohol with the yeast at time zero. Relevant figures for *Schizosaccharomyces* are:

$RGR2 = 0.05\ hour^{-1}$, $ALPF2 = 0.26\ (\%\ alc.)\ (unit\ yeast)^{-1}$

Schizosaccharomyces has the largest alcohol production factor.

22 In this case, *Schizosaccharomyces* would grow relatively slower and *Saccharomyces* faster than suggested by a linear dependency of the reduction factor on the alcohol concentration. In the monocultures, this would not affect the ultimate amount of yeast that is formed, but in the mixture it would lead to less *Schizosaccharomyces* and more *Saccharomyces*. However, the growth curves for the two species in the monoculture would also be of different form. This is not suggested by the data, but the scatter is of course large. The alcohol concentration in the mixture may be calculated by multiplying the final yields with the respective alcohol production factors. This yields a concentration of 1.43%, rather than 1.5%. Hence, there is less yeast in the mixed culture than would be expected. This may be an experimental error.

23 The yeast will grow and increase its amount, and thereby its growth rate and alcohol production rate, until the alcohol concentration approaches 1.5%.
In this situation an infinite amount of yeast will maintain a growth rate which is just sufficient to produce the alcohol that is continuously removed by washing. The removal rate of alcohol is 1.5/10, and the absolute growth rates are obtained by division with the alcohol production factor. This rather ridiculous result is a consequence of the assumption that the maintenance of yeast cells does not need energy and thus does not result in some alcohol production. Obviously, a simulation program which is satisfactory in some situation is not satisfactory in others because simplifications that apply in one situation do not necessarily apply in another.

24 YM equals MALC/ALPF. Since ALPF was calculated from YM, it is not surprising that the YM equals 13 and 5.8 for the species. The first derivative of c/v equals $(-c/v^2) \times (dv/dT)$, when c is a constant, so that the first derivative of Eqn (3.7) is:

$$\frac{dY}{dT} = \frac{-YM}{(1+K \times e^{-RGR \times T})^2} \times (-K \times RGR \times e^{-RGR \times T})$$

The two minus signs cancel, and part of the expression can be replaced by Y itself:

$$\frac{dY}{dT} = Y \times \frac{K \times RGR \times e^{-RGR \times T}}{(1+K \times e^{-RGR \times T})}$$

The fraction $\frac{K \times e^{-RGR \times T}}{(1+K \times e^{-RGR \times T})}$ can also be written as

$$1 - \frac{1}{(1+K \times e^{-RGR \times T})}$$

Substituting Y for a second time gives

$$\frac{dY}{dT} = RGR \times Y \times (1 - Y/YM)$$

In this way the differential equation (3.6) is again arrived at. The initial amount of yeast can be found by substituting for time the value zero into the integrated equation (3.7). This gives:

$$IY = \frac{YM}{1+K}$$

The differential equation for the rate of alcohol production can only be replaced by the integral equation for the amount of alcohol if the initial amounts of yeast that are added are small. If this is not the case, the appropriate amount of alcohol, ALPF × IY, has to be added together with the initial amount of yeast if the analytical solution is to be used. Such restrictions do not hold in the simulation program because no equations are eliminated there.

25 If we again neglect the initial amounts of yeast, the amount of alcohol in the mixed culture is given by

$$ALC = ALPF1 \times Y1 + ALPF2 \times Y2$$

Assume that RED = ALC/MALC for both species, then the growth rates may be formulated as

$$\frac{dY1}{dT} = Y1 \times RGR1 \times \left(1 - \frac{ALPF1 \times Y1}{MALC} - \frac{ALPF2 \times Y2}{MALC}\right)$$

$$\frac{dY2}{dT} = Y2 \times RGR2 \times \left(1 - \frac{ALPF1 \times Y1}{MALC} - \frac{ALPF2 \times Y2}{MALC}\right)$$

R1 and R2 correspond to RGR1 and RGR2, A1 and A2 are equal to ALPF1/MALC and B1 and B2 to ALPF2/MALC. A1 and A2 are equal because it is assumed that Y1 and Y2 are equally sensitive to the alcohol produced by Y1. If Y1 produces some product that is more harmful for Y2 than for Y1, A2 is larger than A1.

8.4 Interference of plants

26 The monocultures stay monocultures. When in the mixed culture half of the area is allotted to the first and half to the second species, the relative seed densities are $n \times 0.5/n = 0.5$ and $m \times 0.5/m = 0.5$. One may also consider n seeds of one species as one seed unit and m seeds of the other as another seed unit.

27 The nominator $k_{12}z_1+z_2$ cancels by division, so that Eqn (4.4) is obtained. Addition of O_1/M_1 and O_2/M_2 gives $(k_{12}z_1+z_2)/(k_{12}z_1+z_2)=1$ so that Eqn (4.3) is obtained.
The results of the calculations are:

z_2(oats)	0.2	0.4	0.6	0.8
α_{12}	1.51	2.09	2.06	2.00
k_{12}	2.05	2.83	2.79	2.71
RYT	1.04	1.02	1.01	0.97

Obviously some smoothing is necessary to obtain a RYT equal to 1, and a constant relative crowding coefficient.
Oats has the highest yield in monoculture, but barley gains in competition.

28 Replacing z_1 and z_2 by Z_1/Z_m and Z_2/Z_m, taking into account that $Z_1+Z_2=Z_m$ and omitting the subscript 1 transforms Eqn (4.5) into:

$$O=\frac{k\times Z/Z_m}{k\times Z/Z_m+(Z_m-Z)/Z_m}M=\frac{k\times Z}{(k-1)\times Z+Z_m}M$$

This equation must be equal to

$$O=\frac{B\times Z}{B\times Z+1}O_m$$

Both equations describe a hyperbola. If the horizontal asymptote and the slope at the origin are equal, the hyperbolae are identical. This means that

$$\frac{k}{k-1}M=O_m$$

and

$$\frac{k}{Z_m}M=B\times O_m$$

The first equation gives directly the expression for O_m, and substitution into the second equation gives

$$B=(k-1)/Z_m$$

29

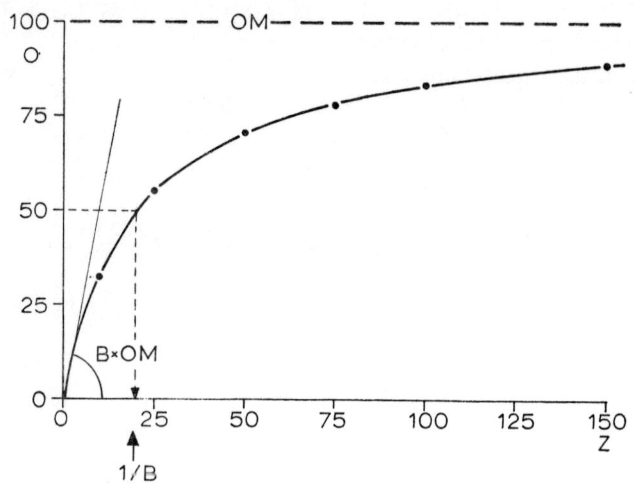

Fig. 37

$$\lim_{Z \to \infty} (O) = O_m$$

$$\lim_{Z \to 0} (O/Z) = \lim_{Z \to 0} \frac{B}{B \times Z + 1} O_m = B \times O_m$$

$$\lim_{Z \to 0} (O/O_m) = \lim_{Z \to 0} \frac{B \times Z}{B \times Z + 1} = B \times Z$$

$$\lim_{Z \to \infty} (O/O_m) = \lim_{Z \to \infty} \frac{B \times Z}{B \times Z + 1} = 1$$

30

	Barley		Oats	
	O_m	B	O_m	B
7 June	470.	0.083	761.	0.0297
21 June	612.	0.574	552.	0.346
5 July	780.	0.778	724.	0.571
19 July	1132.	0.778	956.	1.17

B has the dimension m row^{-1}, because the seed density has the dimension row m^{-1}. The graph for O_m has an unexpected form. The calculated values of O_m (g m^{-2}) are very inaccurate for the first harvest because the yields are very far from their maximum at both densities.
When linearized the growth rate of O_m for barley and oats are 16.4 and 14.5 g m^{-2} day^{-1}, so that the estimated values for O_m at the first harvest are 377 and 333 g m^{-2}. The values of B calculated on this basis are 0.11 and 0.076 m row^{-1}.

31

$$\frac{d(RS)}{dt} = \frac{d(B \times Z)/dt \times (B \times Z+1) - d(B \times Z+1)/dt \times B \times Z}{(B \times Z+1)^2} =$$

$$= \frac{Z \times \frac{dB}{dt} \times B \times Z + Z \times \frac{dB}{dt} - Z \times \frac{dB}{dt} \times B \times Z}{(B \times Z+1)^2} =$$

$$= \frac{B \times Z}{(B \times Z+1)} \times \frac{1}{(B \times Z+1)} \times \frac{1}{B} \times \frac{dB}{dt} =$$

$$= RS \times (1.-RS) \times \frac{1}{B} \times \frac{dB}{dt}$$

The dimension of (dB/dT)/B is $time^{-1}$, the same as a relative growth rate. Usually its value decreases with time. There is an exponential growth when this ratio is constant.

32

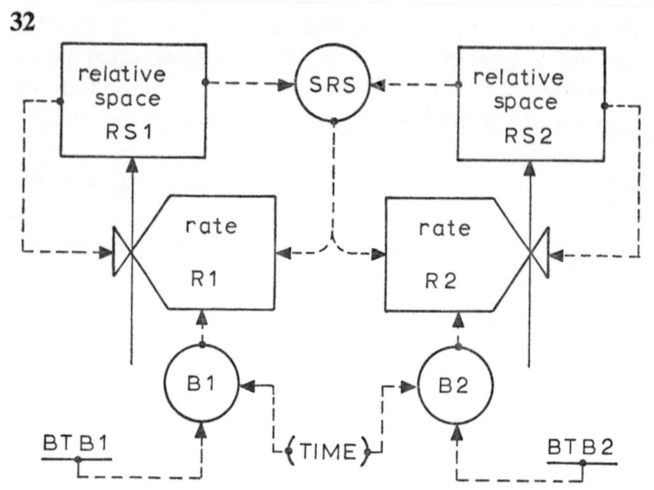

Fig. 38

33 The initial relative growth rate becomes infinite, if B is zero at emergence. The derivative of a variable with respect to time is calculated as the difference between the present value and the value either a sufficiently small time-interval earlier or one ahead, divided by this time-interval, taking care for the sign. In simulation only the first method can be used, as future values are not known. At time zero, however, there is not a past either so that the initial value must be given by the programmer on an `INCON` card.
The use of the derivative function `DERIV` is only allowed, if the derivative is taken of an externally given variable, such as an `AFGEN` function of `TIME`. In this situation the simulation program is used to convert some given variable of time to its derivative with respect to time. As soon as some rate of an integral depends on a derivative of a variable which depends also on some integral, the `DERIV` function must not be used. It produces nonsense results, because an internal, algebraic loop is introduced. When the self-adapting integration method of Runge–Kutta is used, the time-interval will be chosen so small that the choice of the initial value has hardly any influence. With `METHOD RECT`, it is better to initialize properly. We calculated the following initial values of the derivative of B:
Barley: 0.0047 m row^{-1} day^{-1}
Oats : 0.0033 m row^{-1} day^{-1}

The initial values of RS are calculated with

$$RS = \frac{B \times Z}{B \times Z + 1}$$

with $Z = 2$ rows m^{-2} and $B = 0.001$ m^2 row^{-1}.
The simulated results in g m^{-2} are:

Date of harvest	Barley	Oats
7 June	60.5	37.1
21 June	242	140
5 July	338	214
19 July	490	338

These are in good agreement with the experimental results of Table 2.

34

```
TITLE FOUR COMPETING YEAST SPECIES
INITIAL
INCON YI'1,4'=0.1,0.1,0.1,0.1
DYNAMIC
Y'1,4'=INTGRL(YI'1,4',RY'1,4')
RY'1,4'=RGR'1,4'*Y'1,4'*(1.-RED'1,4')
PARAMETER RGR'1,4'=0.2,0.3,0.4,0.5
RED'1,4'=AFGEN(RDTB'1,4',ALC)
FUNCTION RDTB1=0.,0., 1.5,1.
FUNCTION RDTB2=0.,0., 1.,0.8,2.,1.
FUNCTION RDTB3=0.,0., 0.9,1.
FUNCTION RDTB4=0.,0.,1.,0.5,1.5,1.
ALC=INTGRL(0.,ALCP1+ALCP2+ALCP3+ALCP4)
ALCP'1,4'=ALPF'1,4'*RY'1,4'
PARAMETER ALPF'1,4'=0.5,0.4,0.3,0.2
TIMER FINTIM=50.,PRDEL=1.,OUTDEL=1.
PRINT Y'1,4',ALC
END
STOP
```

35 The main differences between the MACRO and the INDEX features are simply of a practical nature. The index feature distinguishes the variables by a number at the end. Since a variable name may consist at the most of 6 alpha-numerical symbols, ABCDE'1,10' creates also the variable ABCDE10 and this is an unacceptably long name. The INDEX feature can be used to any single expression that is normally used in the program. The MACRO feature is unsuitable for this purpose, because every time a MACRO is used, the invoking sentence has to be written. Hence, the only expression in a 'one line macro' may be as well written directly with the proper symbols. The MACRO feature is therefore in general used only when the MACRO definition contains more than one structural statement.

36 A SUBROUTINE contains an algorithm to compute output variables for a main program from input variables out of the main program, any time it is called upon. A MACRO, however, does not contain an algorithm, but is an order to write a part of a CSMP program with the proper symbols every time it is called upon. The different statements written by the MACRO are then sorted in their appropriate places in the FORTRAN-subroutine 'UPDATE'. The individual expression within the original MACRO may be found scattered throughout this subroutine.

37 The correct expression is

$$\mathrm{RSI} = \frac{\mathrm{BI}}{\mathrm{BI+DIST}}$$

in which distance is the inverse of the density of the rows and expressed in m row^{-1}. The initial value BI is close to zero and negligible compared with DIST, so that the expression is sufficiently approximated by RSI = BI/DIST. It can also be said that at initialization the plants in the row are still so small that they do not interfere with plants in other rows.
DBI is the initial value of the derivative and equals approximately $(B_{t+\Delta t} - B_t)/\Delta t$ in which Δt may be chosen as one day because the function for B is linear over this small time-increment.

```
TITLE COMPETITION BETWEEN THREE BARLEY VARIETIES USING THE MACRO FEATURE
MACRO RSI,DBI=BEGIN(BTB,DIST)
      RSI=BI/DIST
      BI=AFGEN(BTB,0.)
      DBI=AFGEN(BTB,1.)-BI
ENDMAC
MACRO O,RS=GROWTH(RSI,DBI,BTB,OMTB)
      RS=INTGRL(RSI,(DB/B)*RS*(1.-SRS))
      B=AFGEN(BTB,TIME)
      DB=DERIV(DBI,B)
      O=RS*AFGEN(OMTB,TIME)
ENDMAC
INITIAL
      RSI1,DBI1=BEGIN(BTB1,DIST1)
      RSI2,DBI2=BEGIN(BTB2,DIST2)
      RSI3,DBI3=BEGIN(BTB3,DIST3)
DYNAMIC
      O1,RS1=GROWTH(RSI1,DBI1,BTB1,OMTB1)
      O2,RS2=GROWTH(RSI2,DBI2,BTB2,OMTB2)
      O3,RS3=GROWTH(RSI3,DBI3,BTB3,OMTB3)
PARAM DIST1=1.2,DIST2=1.2,DIST3=1.2
      SRS=RS1+RS2+RS3
FUNCTION BTB1=(0.,0.001),(30.,.03),(35.,.05),(40.,.09),(45.,.16),(50.,.26),...
      (55.,.38),(60.,.58),(65.,.88),(70.,1.02)
FUNCTION BTB2=(0.,0.0005),(30.,.015),(35.,.025),(40.,.045),(45.,.08),...
      (50.,.13),(55.,.19),(60.,.29),(65.,.44),(70.,.51)
FUNCTION BTB3=(0.,0.001),(30.,.03),(35.,.05),(40.,.09),(45.,.16),(50.,.26),...
      (55.,.38),(60.,.58),(65.,.88),(70.,1.02)
FUNCTION OMTB1=(0.,0.),(70.,5600.)
FUNCTION OMTB2=(0.,0.),(70.,5600.)
FUNCTION OMTB3=(0.,0.),(70.,2800.)
TIMER FINTIM=70.,PRDEL=5.,OUTDEL=5.
PRINT O1,O2,O3,RS1,RS2,RS3
PRTPLT RS1
PRTPLT RS2
PRTPLT RS3
END
STOP
```

```
TITLE COMPETITION BETWEEN THREE BARLEY VARIETIES USING THE INDEX FEATURE
INITIAL
      RSI'1,3'=BI'1,3'/DIST'1,3'
      BI'1,3'=AFGEN(BTB'1,3',0.)
      DBI'1,3'=AFGEN(BTB'1,3',1.)-BI'1,3'
DYNAMIC
      RS'1,3'=INTGRL(RSI'1,3',(DB'1,3'/B'1,3')*RS'1,3'*(1./SRS))
      B'1,3'=AFGEN(BTB'1,3',TIME)
      DB'1,3'=DERIV(DBI'1,3',B'1,3')
      O'1,3'=RS'1,3'*AFGEN(OMTB'1,3',TIME
PARAM DIST'1,3'=1.2,1.2,1.2
      SRS=RS1+RS2+RS3
FUNCTION BTB1=(0.,0.001),(30.,.03),(35.,.05),(40.,.09),(45.,.16),(50.,.26),...
      (55.,.38), (60.,.58), (65.,.88), (70.,1.02)
FUNCTION BTB2=(0.,0.0005),(30.,.015),(35.,.025),(40.,.045),(45.,.08),...
      (50.,.13),(55.,.19),(60.,.29),(65.,.44),(70.,.51)
FUNCTION BTB3=(0.,0.001),(0.,.03),(35.,.05),(40.,.09),(45.,.16),(50.,.26),...
      (55.,.38),(60.,.58),(65.,.88),(70.,1.02)
FUNCTION OMTB1=(0.,0.),(70.,5600.)
FUNCTION OMTB2=(0.,0.),(70.,5600.)
FUNCTION OMTB3=(0.,0.),(70.,2800.)
TIMER FINTIM=70.,PRDEL=5.,OUTDEL=5.
PRINT O'1,3',RS'1,3'
PRTPLT RS'1,3'
END
STOP
ENDJOB
```

8.5 Growth and competition of Paramecium

39 Thorough stirring does not result in a uniform distribution of the Paramecium throughout the liquid medium, but in a random distribution. One-tenth of the solution does therefore contain sometimes more or some times less than exactly one-tenth of the number of protozoa.

40 ....

41

Variable	Dimension	Type of 'variable'
H	protozoon (prot.)	state
AΓOOD	loop	statc
TIME	day	state
CONVF	prot. loop^{-1}	param.
RDR	day^{-1}	param.
FOOD	loop. volume^{-1}	auxil.
RSW	volume.prot.$^{-1}$.day^{-1}	param.
MRDIG	loop.prot.$^{-1}$.day^{-1}	param.
CNRT	loop.day	rate

The example of the relational diagram in Fig. 39 is as schematic as possible.
A simplified integral statement for the net growth rate is:

$$H = \text{INTGRL}(\text{IH}, \text{CONVF} \times \text{CNRT} - \text{RDR} \times H)$$

if a relative death rate is accounted for and

$$H = \text{INTGRL}(\text{IH}, \text{CONVF} \times (\text{CNRT} - \text{MNF} \times H))$$

if maintenance is accounted for by a maintenance factor (MNF in loop.day^{-1}.protozoa^{-1}).
The two formulations are the same with

$$\text{RDR} = \text{CONVF} \times \text{MNF}$$

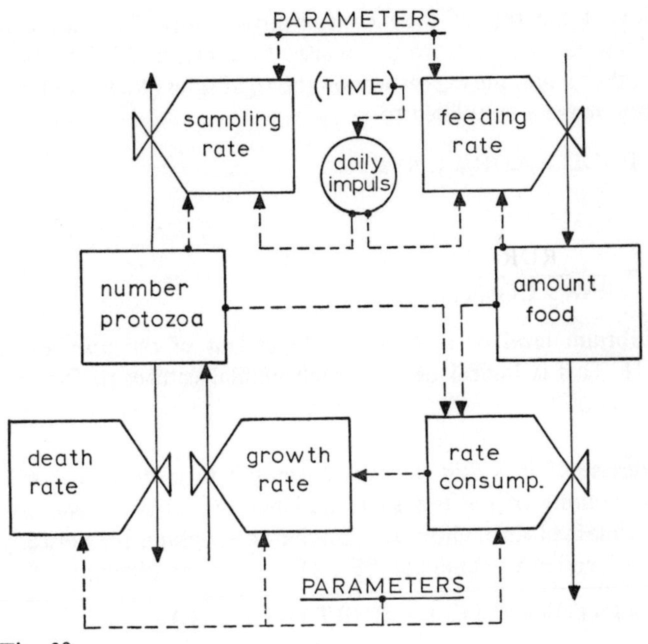

Fig. 39

42 The graph consists out of two line segments: a straight line through the origin and a horizontal 'saturation' level.
MRDIG determines the height of the saturation level. The slope of the line through the origin is determined by RSW, and H is a multiplication factor for the height of the graph as a whole. CNRT equals zero when FOOD is zero. NGR equals zero when

$$CNRT \times CONVF = DR$$

or

$$H \times AMIN1(MRDIG, RSW \times FOOD) \times CONVF = H \times RDR$$

or

$$AMIN1(MRDIG, RSW \times FOOD) = RDR/CONVF$$

The maximum value of the left side of the expression is MRDIG. If the value of the right side is even larger than MRDIG, no value of FOOD exists for which NGR equals zero. In other words, the rate of death could be larger than the rate of growth, even if the animals were eating at the top of their consumptive ability. This leads to rapid extinction, so that there is no need to consider this situation. If, on the other hand, the expression on the right is less than MRDIG, the equation may be simplified to

$$RSW \times FOOD = RDR/CONVF$$

or

$$FOOD = \frac{RDR}{RSW \times CONVF}$$

This equilibrium level of FOOD is independent of the number of protozoa H. This is logical because each animal catches its food on its own.

43 In Exercise 41 it is said that the basic unit to measure volumes will be the volume of the test tube, and not cm^3. This convention must be maintained throughout the simulation program. Numerically there is no objection to eliminate AFOOD and to say directly:

FOOD = INTGRL(L,FEED – CNRTA – CNRTC)

However, the dimensions of the variables are then not consistent: FOOD sometimes means amount of food, as here in the integral, and sometimes it means density of food, as in the expression for CNRT.

44 In such an integration method the criterion for the size of DELT is an error which decreases with the size of DELT. In the considered situation the rate itself is inversely proportional to the size of DELT, so that the calculated error, which is proportional to the product of DELT and the rate, will not decrease when DELT is diminished. Once the occurring error exceeds the error bound, the integration routine starts to halve the value of DELT. As this does not improve the situation, this halving goes on until a minimum allowed value of DELT, called DELMIN, is reached and the simulation is terminated.

45 Four parameters must be estimated for each species. If the order of magnitude of none of the parameters is known one may start by estimating four values for each parameter of which the largest is

10 000 times the smallest. This leads in a first evaluation already to $4^4 = 256$ simulation runs.

46 Since the data scatter considerably your estimates may differ from ours.

	P. aurelia		*P. caudatum*	
	0.5 loop	1 loop	0.5 loop	1 loop
H_{eq}	2500	4500	600	1290
$GR_{.75}$	350	550	100	230
RGR	1.23	1.23	0.74	1.05
CONVF	2800	2200	800	920
RDR	0.46	0.39	0.57	0.61
MRDIG	0.00064	0.00078	0.00176	0.00191
RSW	0.00128	0.00156	0.00176	0.00191

Averaging of the values for 0.5 and 1 loop, and taking into account that *P. caudatum* is about 4 times larger than *P. aurelia*, leads to the following estimates of the parameters:

	P. aurelia	*P. caudatum*
CONVF	3000	750
RDR	0.43	0.59
MRDIG	0.0007	0.0028
RSW	0.0014	0.0018

Some of the reasons why these estimates may be considerably in error are:

a the scatter of the data;

b the population size in the end is not always at H_{eq}, but varies throughout the day. This affects the validity of the two equations with H_{eq} in it;

c it is not certain whether in the beginning the saturation density of food is reached or not. The assumption that it is not so leads to a set of equations in which MRDIG is larger than RSW and which provides different values for RSW;

d in the beginning the density of food varies also throughout the day, so that growth is not exactly exponential;

e the whole concept, vizualized in the simulation program may be wrong.

47 Other parameters being equal, a difference in the relative death rate of 10% causes a difference of only about 8% in the maximum size of the populations for the monoculture. In the mixed culture, however, it is just this 10% difference that makes for survival of one species and extinction of the other; after several days the difference in the population size is much larger than 8%. The same argument holds for the other parameters. Our best estimates are given in Table 4. The program is practically completely presented in the text, so that finalizing it, should not give any difficulties.

48 The food consumption (CNRT) is proportional to the number of protozoa (H) and to the minimum of the maximum rate of digestion (MRDIG) or the food concentration times the rate of searching (FOOD × RSW). The food consumption is proportional to H, when FOOD exceeds MRDIG/RSW or when FOOD is constant. FOOD is constant in the beginning, because the consumption is small compared with supply. Since the death rate is also proportional to H, exponential growth results at the beginning of the experiment.
At the end of the experiment the population grows very fast just after feeding. As soon as the food level is below RDR/(RSW × CONVF), the death rate is larger than the growth rate (Exercise 42). When the food is depleted below this critical level, the population size goes through a maximum, and will be smaller at the end of the day than some time earlier.
The relative death rate is about 0.45 day^{-1} and the relative sampling rate approximately 0.1 day^{-1}, so that death through natural causes is far larger than through sampling. It is interesting to remark that Gause did not only discard the sampled amount for practical reasons, but also because he was (unnecessarily) afraid that the natural death rate would be so small that one species would not replace the other in competition.

49 According to the Poisson distribution function, the standard deviation is the square root of the sampled number so that the relative standard deviation is inversely proportional to this square root. The population of *P. aurelia* is about 4 times larger than of *P. caudatum*, so that its relative standard deviation is about half, as reflected in the scatter of the observations.

8.6 Modelling of development, dispersion and diffusion

50 a At 14°C the function DVRTB is interpolated between the points (12.,0.) and (26.,0.035); this gives 0.005 day^{-1} for DVR. Accordingly, the development stage at 20 days is 0.1.
b DVR equals 0 at 7 °C and 0.0225 day^{-1} at 21 °C. The development stage at 20 days is $20 \times 0.5 \times 0. + 20 \times 0.5 \times 0.0225$ or 0.225.
In both situation a and b, the average temperature is 14 °C, but development is much faster in b because of the variation in the temperature, combined with a more than linear temperature response.
c At 30 °C the rate of development is 0.039 day^{-1}, so that at 20 days a development stage of 0.78 is reached.
d The temperature is 40 °C for 6 hours, so that DVR equals 0.041. The temperature is 26.7 °C with a DVR of 0.0357 for the other 18 hours. At 20 days, DVS equals $20 \times 0.25 \times 0.041 + 20 \times 0.75 \times 0.0357$ or 0.7405. This is slightly less compared to situation c, although the average temperature was the same during this period. Here the variability has caused a decrease in development, because of the less than linear temperature response in this region.

51 With a constant relative death rate RDR, and no birth, the population as a function of time is

$$H = HI \times e^{-RDR \times TIME}$$

as derived in Section 2.1. When RDR has the dimension $year^{-1}$, the total death during the first year amounts to

$$HI \times (1. - e^{-RDR})$$

so that the relative death rate on a yearly basis equals

$$1. - e^{-RDR}$$

For small values of RDR this approaches RDR. When RDR is 0.02 the error is only 1%.
When RDR is zero and the relative birth rate equals RBR, the size of the population is

$$H = HI \times e^{RBR}$$

after one year.
The birth rate on a yearly basis is then

$e^{RBR} - 1$.

The integration routine METHOD RECT must be used because a division by DELT occurs in the expression for the rates. This was also the case in the example of the paramecia.

52 The data of set 1 must be used. The time-interval of integration is a half year because the IMPULS function works at 2.5, 7.5 etc. years. The birth and death rate data can also be used when time-intervals smaller than one year are applied but for large time-intervals they have to be recalculated on that basis. It is a good custom in the Netherlands to maintain graves for a limited period of time. Therefore, the question as to the number of graves may be relevant. To calculate the number of graves another series of at least 10 classes of 5 years must be introduced. The birth rate of the graves equals the death rate of the population and the 'relative death rate' of graves is zero.
Demographically, there is hardly any difference between death occurring during the first years of life and a decrease in birth rate corresponding with the death rate in excess of the 'normal' death rate during the first year. However, if this correction is made, it must be taken into account that the chances of dying during the first year are not the same for boys and girls, so that the sex ratio has to be corrected accordingly. The total male and female population after 50 years equals 10.49×10^6 and 10.39×10^6, respectively and the total number of graves are 1.853×10^6, 4.340×10^6 and 7.528×10^6 when maintained for 10, 25 and 50 years. The simulation program for the growth of the population is given in Fig. 40.
The simulation of the number of graves is programmed in exercise 61.

The birth rates and death rates per thousand are recalculated on a relative basis in a NOSORT section because statements of the form:

```
MRDR1 = MRDR1 * ...............
```

cannot be sorted (Section 2.3).
It would have been also possible to rename the variables.

53 GS = 1 means that all seeds are germinated. Simulation beyond this point does not make sense, as far as germination is concerned.

```
PARAM WRDR'1,19'=11.4,1.2,.3,.3,.4,.4,.6,1.,1.5,2.5,4.,5.5,8.,13.,   ...
      20.,50.,120.,250.,500.
PARAM MRDR'1,19'=15.6,1.8,.5,.5,.7,1.,1.2,1.5,2.2,4.,6.5,9.,11.5,16.,...
      35.,70.,150.,300.,600.
INITIAL
NOSORT
*     CONVERSION OF DEATHS PER THOUSAND PER YEAR TO RELATIVE DEATH RATES
      MRDR'1,19'=0.001*MRDR'1,19'
      WRDR'1,19'=0.001*WRDR'1,19'
SORT
PARAM RBR'1,16'=0.,0.,0.,0.,.091,.159,.152,.084,.036,.01,0.,0.,0.,0.,...
      0.,0.
INCON WI'1,19'=291000.,584000.,570000.,548000.,548000.,487000.,      ...
      400000.,380000.,379000.,377000.,353000.,327000.,310000.,       ...
      262000.,226000.,180000.,110000.,50000.,33000.
INCON MI'1,19'=305000.,612000.,597000.,575000.,576000.,517000.,      ...
      429000.,399000.,382000.,367000.,338000.,306000.,280000.,       ...
      223000.,184000.,150000.,90000.,40000.,13000.
*
DYNAMIC
      W'2,19'=INTGRL(WI'2,19',WFL'1,18'-WFL'2,19'-W'2,19'*WRDR'2,19')
      M'2,19'=INTGRL(MI'2,19',MFL'1,18'-MFL'2,19'-M'2,19'*MRDR'2.19')
      M1=INTGRL(MI1,MBR-M1*MRDR1-MFL1)
      W1=INTGRL(WI1,WBR-W1*WRDR1-WFL1)
      BR'1,16'=W'1,16'*RBR'1,16'
      TBR=BR1+BR2+BR3+BR4+BR5+BR6+BR7+BR8+BR9+BR10+BR11+BR12
      WBR=TBR/2.048
      MBR=TBR*1.048/2.048
      WFL'1,19'=PUSH*W'1,19'/DELT-WRDR'1,19'*W'1,19'
      MFL'1,19'=PUSH*M'1,19'/DELT-MRDR'1,19'*M'1,19'
      PUSH=IMPULS(2.5,5.)
      TW=W1+W2+W3+W4+W5+W6+W7+W8+W9+W10+W11+W12+W13+W14+W15+W16+W17  ...
      +W18+W19
      TM=M1+M2+M3+M4+M5+M6+M7+M8+M9+M10+M11+M12+M13+M14+M15+M16      ...
      +M17+M18+M19
      TP=TM+TW
METHOD RECT
TIMER FINTIM=50.,DELT=.5,OUTDEL=5.,PRDEL=5.
PRINT TP,TM,TW,TBR,M'1,19',W'1,19'
PRTPLT TP,TBR
END
STOP
ENDJOB
```

Fig. 40

The termination of the program is achieved by inserting a finish card:

```
FINISH GS=1.
```

54 It is a rather arbitrary choice to give the value 1 to the stage of germination; it could just as well be 1000. Whatever the value, it is passed going through N classes. Accordingly each class covers the chosen germination value divided by N.

55 The simulation program may read as follows.

```
PARAMETER N=10
H1=INTGRL(1000.,-H1*PUSH/DELT)
H'2,10'=INTGRL(0.,(H'1,9'-H'2,10')*...
  PUSH/DELT)
PUSH=INSW(GS-1/N,0.,1.)
GS=INTGRL(0,VDV-PUSH/(N*DELT))
PARAMETER VDV=0.143
METHOD RECT
```

etc.
The `METHOD RECT` must be used for integration because of the discontinuous changes.
The seeds germinate at 1./0.143 = 7.0 days.

56 The average germination period is not necessarily the moment when 50% is germinated. When G is the rate of germination, the mathematical definition of the average germination period is:

$$\mathrm{AGP} = \frac{\int_0^\infty G \times T \times dT}{\int_0^\infty G \times dT}$$

To execute this calculation, the curve is divided in sections of 1 day and the formula

$$\mathrm{AGP} = \sum_{n=0}^{27} G_n \times T_n \Big/ \sum_{n=0}^{27} G_n$$

is used, in which n is the number of the day.
Because the cumulative curve adds up to 100%, the denominator in

this formula is 100. AGP appears to be 16.4 day.
Similarly, the formula for the variance (the square of the standard deviation) is:

$$\mathrm{VAR} = \sum_{n=0}^{27} G_n \times (T_n - \mathrm{AGP})^2 \Big/ \sum_{n=0}^{27} G_n$$

This value is 11.91, so that the standard deviation is 3.45 day.

57 It is assumed that germination on the nth day means germination at the beginning of the nth day. The average germination periods in days are then: $100 \times 5/100 = 5$, $100 \times 10/100 = 10$, $(50 \times 5 + 50 \times 10)/100 = 7.5$, $(75 \times 5 + 25 \times 10)/100 = 6.25$. If it is assumed that germination occurs during the nth day, 0.5 day must be added to these values. The dimension of the relative germination rate is day^{-1} and of the average germination period is days. The product is therefore dimensionless. Its value appears to be approximately one.
The average germination period is

$$\mathrm{AGP} = -\frac{1}{\mathrm{HI}} \int_0^{\infty} T \times \frac{dH}{dT} \times dT = \frac{1}{\mathrm{HI}} \int_0^{\mathrm{HI}} T \times dH$$

The integral is the surface below the H versus T curve, so that the expression can be replaced by

$$\mathrm{AGP} = \frac{1}{\mathrm{HI}} \int_0^{\infty} H \times dT = \frac{1}{\mathrm{HI}} \int_0^{\infty} \mathrm{HI} \times e^{-\mathrm{RDV} \times T} \times dT =$$

$$= -\frac{1}{\mathrm{RDV}} \times \left[e^{-\mathrm{RDV} \times T} \right]_0^{\infty} = \frac{1}{\mathrm{RDV}}$$

Hence $\mathrm{RDV} \times \mathrm{AGP} = 1$.

58 The time constant of the system is found by reducing the expression for the rate to

$$(dH/dT)/H$$

This results here in 1/REST, so that TAU = REST.

59 According to Eqn (6.2), $(1 - F)$ equals 25/16, so that F should be negative. Obviously the number of classes that is chosen is too large to obtain a relative dispersion of 0.25. With N = 16 and F = 0, the proper dispersion is obtained.

60

TIME	H1	H2	H3	H4
0	1	0	0	0
0.5 × REST	0.5	0.5	0	0
1.0 × REST	0.25	0.5	0.25	0
1.5 × REST	0.125	0.375	0.375	0.125

This is the binomial probability distribution function.
B equals TIME/(F × REST) and f equals F, so that f × B = TIME/REST. With f × B small and constant and B increasing, the binomial approaches the Poisson distribution function.
This situation is achieved here when the lowest value of F i.e. DELT/REST is substituted in the expression for B, and DELT approaches zero. Then B equals TIME/DELT and approaches infinity.
The expectation value and the variance of the Poisson distribution are f × B and (1 − f) × f × B or in the other terms TIME/REST and (1 − F) × TIME/REST. To convert the variance in terms of time, rather than class, the expression must be multiplied by $REST^2$. This gives the expression $S^2 = (1 - F) \times TIME \times REST$ for the variance which reduces into Eqn 6.2 when TIME is replaced by AGP. It is also recalled that for high expectation values the Poisson function approaches the Gaussian function with a variance equal to the expectation value of the mean.

61 The age-classes are now indeed 0–5, 5–10, 10–15 etc. so that the set with the class centres at 2.5, 7.5, ... years must be used. The residence time is 5, and the value of F is of course 1 because age-classes advance per definition without dispersion.
The simulation program, inclusive the number of graves is given in Fig. 41.

```
PARAM WRDR'1,19'=4.0,.8,.3,.3,.4,.5,.8,1.2,2.,3.2,4.7,6.7,10.5,       ...
      16.5,35.,85.,180.,380.,760.
PARAM MRDR'1,19'=6.0,.7,.5,.6,.9,1.,1.4,1.8,3.1,5.2,7.8,10.7,13.7,    ...
      25.5,52.,110.,200.,400.,900.
INITIAL
      INVD=1./DELT
NOSORT
      MRDR'1,19'=0.001*MRDR'1,19'
      MRDR'1,19'=0.001*MRDR'1,19'
SORT
PARAM RBR'1,16'=0.,0.,0.,.022,.137,.188,.113,.055,.016,.002,0.,0.,0.,...
      0.,0.,0.
PARAM WI'1,19'=582000.,587000.,553000.,543000.,554000.,420000.,       ...
      380000.,381000.,378000.,367000.,330000.,323000.,298000.,        ...
      226000.,226000.,150000.,70000.,25000.,13000.
PARAM MI'1,19'=611000.,613000.,580000.,569000.,583000.,452000.,       ...
      405000.,393000.,371000.,362000.,314000.,297000.,262000.,        ...
      184000.,184000.,120000.,60000.,20000.,3000.
DYNAMIC
      MO=INTGRL(0.,MBR-DRMO-MO/2.5)
      WO=INTGRL(0.,WBR-DRWO-WO/2.5)
      M1=INTGRL(MI1,MO/2.5-DRM1-MFL1)
      W1=INTGRL(WI1,WO/2.5-DRW1-WFL1)
      BR'1,16'=W'1,16'*RBR'1,16'
      TBR=BR1+BR2+BR3+BR4+BR5+BR6+BR7+BR8+BR9+BR10+BR11+BR12
      WBR=TBR/2.048
      MBR=TBR*1.048/2.048
      W'2,19'=INTGRL(WI'2,19',WFL'1,18'-WFL'2,19'-DRW'2,19')
      M'2,19'=INTGRL(MI'2,19',MFL'1,18'-MFL'2,19'-DRM'2,19')
      WFL'1,19'=PUSH*W'1,19'*INVD-DRW'1,19'
      MFL'1,19'=PUSH*M'1,19'*INVD-DRM'1,19'
      DRW'1,19'=W'1,19'*WRDR'1,19'
      DRWO=WO*WRDR1
      DRMO=MO*MRDR1
      DRM'1,19'=M'1,19'*MRDR'1,19'
      PUSH=IMPULS(2.5,5.)
      TM=M1+M2+M3+M4+M5+M6+M7+M8+M9+M10+M11+M12+M13+M14+M15+M16+MO    ...
      +M17+M18+M19
      TW=W1+W2+W3+W4+W5+W6+W7+W8+W9+W10+W11+W12+W13+W14+W15+W16+WO    ...
      +W17+W18+W19
      TP=TM+TW
      TDRW=DRWO+DRW1+DRW2+DRW3+DRW4+DRW5+DRW6+DRW7+DRW8+DRW9+DRW10+   ...
      DRW11+DRW12+DRW13+DRW14+DRW15+DRW16+DRW17+DRW18+DRW19+WFL19
      TDRM=DRMO+DRM1+DRM2+DRM3+DRM4+DRM5+DRM6+DRM7+DRM8+DRM9+DRM10+   ...
      DRM11+DRM12+DRM13+DRM14+DRM15+DRM16+DRM17+DRM18+DRM19+MFL19
      TDR=TDRM+TDRW
      GO=INTGRL(0.,TDR-GO/2.5)
      G1=INTGRL(0.,GO/2.5-FLG1)
      G'2,10'=INTGRL(0.,FLG'1,9'-FLG'2,10')
      FLG'1,10'=G'1,10'*PUSH*INVD
      TG10=GO+G1+G2
      TG25=TG10+G3+G4+G5
      TG50=TG25+G6+G7+G8+G9+G10
TIMER FINTIM=50.,DELT=.5,OUTDEL=5.
PRTPLT TG10,TG25,TG50,TM,TW
METHOD RECT
PRTPLT TP,TBR
END
STOP
ENDJOB
```

Fig. 41

62 When the topmost layer has the number one, `FLW1` gives a positive contribution and `FLW2` a negative one. This means that the downward direction is taken as positive, because in that case `FLW 1` goes into the first layer. This definition of the sign must be taken into account when the expression for FLW is written.

63 The flow into the first layer is governed by the temperature difference between the surface of the soil and the centre of the first layer. The distance between these levels is only half of the thickness of the compartment.
The unit of time is day, as follows from the definition of TMPS and the remark that a cyclic daily fluctuation is assumed. Every time when the argument of a sine has the value $2\pi(=6.28)$, a full cycle is completed.
TAV represents the average temperature of the soil surface, and TAMPL is the amplitude of the sine wave.

64

Variable	Dimension
TMP,TI,TMPS,TAV,TAMPL	°C
TCOM	cm
VHCAP	$J\ cm^{-3}\ C$
HC,HCI	$J\ cm^{-2}$
COND	$J\ cm^{-1}\ C^{-1}\ day^{-1}$
FLW,NFL	$J\ cm^{-2}\ day^{-1}$

It is extremely important to be aware of the units, as the numerical value of the properties depends on the units. It is recommended to use a consistent set of units: this is done here because the unit of heat (J) is considered as a basic unit. As soon as transformations to other forms of energy occur it is recommended to use the international system with kg, m, s as a basis.

65 The easiest method is to specify the conductivity and the heat capacity as a function of depth in an AFGEN function. A new variable for depth, DPT, must then be introduced, whereby DPT1 equals $0.5 \times$ TCOM, etc. Also TCOM can be varied with the number of the layer. TCOM1, TCOM2, etc. must then be specified.

8.7 Growth and development of Helminthosporium maydis

66 This is very dangerous because the responses to temperature, humidity, wind and probably radiation are likely to be non-linear (Compare with Exercise 50).

67

```
*      WEATHER
       TEMP=AFGEN(TEMPT,HOUR)
       WIND=AFGEN(WINDT,HOUR)
       WET=AFGEN(WETT,HOUR)
       LITE=AFGEN(LITET,HOUR)
       DRY=1.-WET
       RAIN=AFGEN(RAINT,HOUR)
FUNCTION RAINT=0.,0.,24.,0.
FUNCTION TEMPT=0.,14.,12.,35.,24.,14.
FUNCTION WETT=0.,1.,7.99,1.,8.,0.,19.99,0.,20.,1.,24.,1.
FUNCTION WINDT=0.,1.,6.,1.,14.,4.,19.,2.,24.,1.
FUNCTION LITET=0.,-1.,5.99,-1.,6.,1.,20.,1.,20.01,-1.,24.,-1.
FUNCTION LAIT=(0.,3.),(140.,3.)
       LAI=AFGEN(LAIT,TIME)
       HOUR=24.*(TIME-AINT(TIME))
```

68 The expressions are similar because the growth rate of the lesions is linearly dependent on the difference between their maximum area and actual area and because the maximum area of the lesions is the same.

```
*      GROWTH OF LESIONS
       CNL'1,6'=INTGRL(CNLI'1,6',RCNL'1,6'-RCNL'2,7')
       RCNL1=RTN
       CNL7=INTGRL(0.,RCNL7)
       RCNL'2,7'=CNL'1,6'*2.
       AL=INTGRL(0.,RAL)
       RAL=PAL*(MALS*CNL7-AL)
       PAL=AFGEN(PALT,TEMP)
FUNCTION PALT=0.,0.,10.,.14,18.,.33,23.,.8,30.,.8,35.,.14,40.,0.
PARAMETER MALS=1.E-8
```

The total residence time of the lesions in the non-visible stage is $6 \times 0.5 = 3$ days. Note that the contents of each class is multiplied by 2 day^{-1} rather than divided by 0.5 day. This is because multiplication takes considerably less computing time than division and this starts to count with larger programs.
Some lesions are already visible after 1.5 day because of the dispersion during passage through the six compartments. If we assume that DELT is small compared with 0.5 day, the standard deviation of lesion appearance is calculated with Eqn (6.1). The curve for total

lesion area (AL) is sigmoid for two reasons: the dispersed lesion appearance and the proportionality of growth with the difference of maximum and current area.

69

```
*       FORMATION OF GREEN STALKS
        ROP=MOA*RAL
        ROP1=ROP+EMPT1+EMPT2+EMPT3+EMPT4
        OP'1,3'=INTGRL(0.,ROP'1,3'-ROP'2,4'-EMPT'1,3')
        ROP'2,4'=OP'1,3'*16.*WET
        OP4=INTGRL(0.,ROP4-ROX-ROG-EMPT4)
        EMPT'1,4'=DRY*OP'1,4'*INVD
PARAMETER MOA=300.E10
        POG=INSW(LITE,AFGEN(POGD,TEMP),AFGEN(POGL,TEMP))*WET
        POX=INSW(LITE,AFGEN(POXD,TEMP),AFGEN(POXL,TEMP))*WET
FUNCTION POGL=0.,0.,14.,.04,18.,.12,23.,1.4,30.,1.2,35.,0.
FUNCTION POXL=0.,0.,14.,.04,18.,.12,23.,1.4,30.,0.
FUNCTION POGD=0.,0.,14.,.1,18.,.27,23.,.27,30.,1.33,35.,.67,40.,0.
FUNCTION POXD=0.,0.,14.,.02,18.,.03,23.,.18,30.,.88,35.,1.54,40.,0.
        ROG=POG*OP4
        ROX=POX*OP4
```

Note that the division with DELT is replaced by multiplication with INVD, a parameter which is in the initial section defined as 1/DELT. The dimension of ROG equals number of green stalks per ha soil surface per day. POG and POX in the light at 21 °C are

$$0.15 + \frac{21-18}{23-18}(1.4-0.12) = 0.828$$

in both cases.

The fraction of stalks that are formed is therefore 0.828/(0.828 + 0.828) = 0.5. The fraction of stalks that are formed under other conditions is calculated in a similar way.

70 The number of spores after time T equals

$$S_T = S_o \times e^{-RBETR \times T}$$

so that

$$RBETR = -(\ln(S_T/S_o))/T = -(\ln(1-0.63))/5$$
$$= 1/5\ h^{-1} = 4.8\ day^{-1}$$

which may according to the function BEATT, be caused by a rainfall of 16 mm h^{-1}.

71

```
*       FORMATION OF SPORES ON GREEN STALKS
        GST'1,3'=INTGRL(0.,RGST'1,3'-RGST'2,4'-DGST'1,3'-BGST'1,3')
        RGST1=ROG
        RGST'2,3'=GST'1,2'*16.*WET
        DGST'1,3'=DRY*GST'1,3'*INVD
        BGST'1,3'=RBETR*GST'1,3'
        RBETR=AFGEN(BEATT,RAIN)
FUNCTION BEATT=0.,0.,.25,.08,.75,.32,6.25,2.,18.8,5.6,25.,6.7
        RGS=GST3*INSW(LITE,AFGEN(PGSD,TEMP),AFGEN(PGSL,TEMP))*WET
FUNCTION PGSL=0.,0.,14.,.15,18.,1.44,23.,.32,30.,0.,40.,0.
FUNCTION PGSD=0.,0.,14.,.06,18.,14.,23.,14.,30.,.44,40.,0.
        RGST4=RGS
        NGST=GST1+GST2+GST3
```

The dimension of **RGS** is spores on green stalks per hectare soil surface per day.

72

```
*       FORMATION OF SPORES ON DRIED STALKS
        DGST=DGST1+DGST2+DGST3
        DDST=DDST1+DDST2+DDST3+DDST4
        RDST1=(DDST+DGST)+RSR+ROG
        RDST5=RDS
        DST'1,4'=INTGRL(0.,RDST'1,4'-RDST'2,5'-DDST'1,4'-BDST'1,4')
        RDST'2,4'=DST'1,3'*16.*WET
        DDST'1,4'=DRY*DST'1,4'*INVD
        BDST'1,4'=RBETR*DST'1,4'
        RDS=DST4*INSW(LITE,AFGEN(PDSD,TEMP),AFGEN(PDSL,TEMP))*WET
FUNCTION PDSL=0.,0.,14.,.17,18.,1.75,23.,.25,30.,0.,40.,0.
FUNCTION PDSD=0.,0.,14.,.07,18.,2.95,23.,2.2,30.,.53,35.,0.,40.,0.
        NDST=DST1+DST2+DST3+DST4
```

The rate of spore removal (**RSR**) must still be calculated.

73 Again

$$S_T = S_o \times e^{-RWASH \times T}$$

with T in days. Hence

$$RWASH = -(\ln(1-.86))/(3/24) = 15.7 \text{ day}^{-1}$$

74 $RBLOW = -\ln(1.-.05)/(3/24) = 0.408 \text{ day}^{-1}$

75

```
*       SPORE DISPERSAL
        RSP=RDS+RGS
        STSP=INTGRL(0.,RSP-RSR)
        RSR=SPRR+SPRD
        SPRR=RWASH*STSP
        SPRD=RBLOW*STSP*DRY
        RWASH=RAIN*2.62
        RBLOW=.102*AMAX1(1.,WIND*WIND)
        RASP=SPRR*REFF+SPRD*WEFF
        REFF=AFGEN(REFFT,RAIN)
FUNCTION REFFT=0.,0.2,2.5,0.003,10.,0.003
*       WEFF=LAI*0.01
```

Host exhaustion could be taken into account by reducing the leaf area (LAI) by the area of the lesions (AL). If the disease is so severe that AL is not small compared with LAI, the disease should also feedback on the growth of the crop.

76

```
*       GERMINATION AND PENETRATION
        CSF=INTGRL(0.,RASP-RKSP-RG-REXS-WCSF)
        WCSF=RWASF*CSF
        RG=CSF*AFGEN(PFTT,TEMP)*(1.-KILL)*WET
        REXS=CSF*AFGEN(PFXT,TEMP)*(1.-KILL)*WET
FUNCTION PFTT=0.,0.,10.,.4,15.,1.8,20.,4.6,23.,7.,35.,3.7,40.,0.
FUNCTION PFXT=0.,0.,10.,0.,15.,1.8,20.,4.2,23.,2.6,35.,3.7,40.,0.
        RKSP=KILL*CSF*INVD
        PROCEDURE KILL=DESS(WET)
        KILL=0.
        IF(WETP-WET).GT.0.01) KILL=1.
        WETP=WET
ENDPRO
        CGT=INTGRL(0.,RG-RKGT-RTN-REXT-WCGT)
        WCGT=RWASH*CGT
        RKGT=KILL*CGT*INVD
        REXT=CGT*AFGEN(PTXT,TEMP)*(1.-KILL)*WET
        RTN=CGT*AFGEN(PTNT,TEMP)*(1.-KILL)*WET
FUNCTION PTNT=0.,0.,18.,.48,23.,.65,30.,.25,35.,0.,40.,0.
FUNCTION PTXT=0.,0.,18.,1.3,23.,2.6,30.,2.2,35.,0.,40.,0.
```

Note that RG and some other rates are also multiplied by (1-KILL) to avoid that the same spores or germs are killed upon desiccation and transferred at the same time.

77 Another rate (INVR) has to be added to the integral of the lesions on the foliage (CSF). This rate equals

```
INVR = 1000*1000*24*DRY*INSW(LITE,0,1)...
     *INSW(TIME-7.,1,0)
```

78

```
OUT1=DEBUG(10,0.)
OUT2=DEBUG(2,5.)
OUT3=DEBUG(2,5.5)
```

79

```
TEMP=AVTMP+AMPL*SINE(6.28*(TIME+8/24))
```

80 You are now on your own.

References

Baeumer, K. & C. T. de Wit, 1968. Competitive interference of plant species in monocultures and in mixed stands. Neth. J. agric. Sci. 16: 103–122.

Berger, R. D., 1970. Forecasting *Helminthosporium turcicum* attacks on Florida sweet corn. Phytopathology 60: 1284.

Bergh, J. P. van den, 1968. An analysis of yields of grasses in mixed and pure stands. Versl. Landbouwk. Onderz. (Agr. Res. Rep.) 714, Wageningen.

Brennan, R. D. & M. Y. Silberberg, 1968. The system/360 continuous modeling program. Simulation 11: 301–311.

Donald, C. M., 1963. Competition among crop and pasture plants. Adv. Agronomy 15: 1–118.

Faddeev, D. K. & V. N. Faddeeva, 1964. Computational methods of linear algebra. W. H. Freeman Co., San Francisco.

Forrester, J. W., 1961. Industrial Dynamics. MIT-press, Boston.

Gause, G. F., 1934. The struggle for existence. Williams and Wilkins, Baltimore.

Goudriaan, J., 1973. Dispersion in simulation models of population growth and salt movement in the soil. Neth. J. agric. Sci. 21: 269–281.

Goudriaan, J. & P. E. Waggoner, 1972. Simulating both aerial microclimate and soil temperature from observations above the foliar canopy. Neth. J. agric. Sci. 20: 104–124.

Goudriaan, J. & C. T. de Wit, 1973. A re-interpretation of Gause's population experiments by means of simulation. Anim. Ecol. 42: 521–530.

IBM, 1972. System/360 Continuous System Modeling Program (360A-CX-16X). User's Manual GH20-0367-4 Techn. Publ. Dept., White Plains, U.S.A.

Janssen, J. G. M., 1973. Simulation of germination of winter annuals in relation to microclimate and micro-distribution. Oecologia (in press).

Janssen, J. G. M., 1973. Effect of light, temperature and seed age on the germination of the winter annuals *Veronica arvensis* L. and *Myosotis ramosissima* Rochel ex. Schult. Oecologia 12: 141–146.

Lotka, A. J., 1925. Elements of physical biology. Williams and Wilkins, Baltimore.

Milne, W., 1960. Numerical solution of differential equations. McGraw-Hill, New York.

Royle, D. J., 1973. Quantitative relationships between infection by the hop downy mildew pathogen, *Pseudoperonospora humuli* and weather and inoculum factors. Ann. appl. Biol. 73: 19–30.

Shaner, G. E. et al., 1972. EPIMAY, an evaluation of a plant disease display model. Bull. Agr. Exp. Sta. Purdue Univ., West Lafayetta, Indiana, U.S.A.

Unesco, 1972. Expert Panel on the role of systems analyses and modeling approaches in the programme on man and biosphere. MAB report series No. 2, Paris.

Volterra, V., 1931. Variations and fluctuations of the number of individuals in animal species living together. In R. N. Chapman, 'Animal Ecology', McGraw-Hill, New York.

Waggoner, P. E., J. G. Horsfall & R. J. Lukens, 1972. EPIMAY, a simulator of southern corn leaf blight. Bull. Conn. Agr. Exp. Sta. New Haven, Conn., U.S.A.

Wit, C. T. de, 1960. On competition. Versl. Landbouwk. Onderz. (Agric. Res. Rep.) 66.8, Wageningen.

Wit, C. T. de, R. Brouwer & F. W. T. Penning de Vries, 1970. The simulation of photosynthetic systems. In: Prediction and Measurement of Photosynthetic Productivity. Proc. of the IBP/PP Technical Meeting, Třeboň, Sept. 1969, 47–70.

Wit, C. T. de, P. G. Tow & G. C. Ennik, 1966. Competition between legumes and grasses. Versl. Landbouwk. Onderz. (Agric. Res. Rep.) 687, Wageningen.